CENTENNIAL HISTORY OF THE
CARNEGIE INSTITUTION OF WASHINGTON

Volume V

THE DEPARTMENT OF EMBRYOLOGY

---

edited by

JANE MAIENSCHEIN, MARIE GLITZ AND
GARLAND E. ALLEN

CAMBRIDGE
UNIVERSITY PRESS

CAMBRIDGE UNIVERSITY PRESS
Cambridge, New York, Melbourne, Madrid, Cape Town,
Singapore, São Paulo, Delhi, Mexico City

Cambridge University Press
The Edinburgh Building, Cambridge CB2 8RU, UK

Published in the United States of America by Cambridge University Press, New York

www.cambridge.org
Information on this title: www.cambridge.org/9781107412422

© Carnegie Institution of Washington 2004

First published 2004
First paperback edition 2012

A catalogue record for this publication is available from the British Library

ISBN 978-0-521-83082-9 Hardback
ISBN 978-1-107-41242-2 Paperback

Centennial History of the Carnegie Institution of Washington
Volume V   The Department of Embryology

Founded in 1914, the Department of Embryology of the Carnegie Institution of Washington has made an unparalleled contribution to the biological understanding of embryos and their development. Originally much of the research was carried out through experimental embryology, but by the second half of the twentieth century, tissue and cell cultures were providing histological information about development, and biochemistry and molecular genetics have now taken center stage. This final volume in a series of five histories of the Carnegie Institution of Washington provides a history of embryology and reproductive biology spanning a hundred years. It provides important insights into the evolution of both scientific ideas and the public perception of embryo research, concluding with a reflection on current debates.

JANE MAIENSCHEIN is Regents' Professor and Parents Association Professor at Arizona State University, and Director of the Center for Biology and Society at ASU. She specializes in the history and philosophy of biology and the way that biology, bioethics, and biopolicy play out in society.

MARIE GLITZ is the Managing Editor of the *Journal of the History of Biology*. Her interests, stemming from her interdisciplinary science degree, focus on the history and philosophy of biology and medicine as well as current biomedical issues.

GARLAND E. ALLEN is Professor of Biology at Washington University. His research interests are in the area of history and philosophy of biology – particularly genetics, embryology, and evolution – and their interrelationships between 1880 and 1950.

CONTENTS

CONTRIBUTORS

Garland E. Allen, Department of Biology, Washington University, St Louis, MO 63130, USA

Donald D. Brown, Department of Embryology, Carnegie Institution of Washington, 115 West University Parkway, Baltimore MD 21210, USA

Adele E. Clarke, Department of Social and Behavioral Sciences, Box 0612, 3333 California St, Suite 455, University of California, San Francisco, CA 94143-0612, USA

Marie Glitz, School of Life Sciences, Arizona State University, PO Box 874501, Tempe, AZ 85284-4501, USA

Elizabeth Hanson, The Rockefeller University, 1230 York Avenue, Box 257, New York, NY 10021, USA

Hannah Landecker, Department of Anthropology, Rice University MS20, PO Box 1892, Houston, TX 77251-1892, USA

Jane Maienschein, School of Life Sciences, Arizona State University, PO Box 874501, Tempe, AZ 85287-4501, USA

Adrianne Noe, National Museum of Health and Medicine, Armed Forces Institute of Pathology, Washington, DC 20306, USA

Allan Spradling, Department of Embryology, Carnegie Institution of Washington, 115 West University Parkway, Baltimore, MD 21210, USA

FOREWORD

In 1902 Andrew Carnegie, a steel magnate turned philanthropist, had a brilliant idea. Carnegie was prescient in recognizing the important role that science could play in the advancement of humankind. He also believed that the best science came by providing "exceptional" individuals with the resources they need in an environment that is free of needless constraints. He created the Carnegie Institution as a means to realize these understandings, directing the Institution to undertake "projects of broad scope that may lead to the discovery and utilization of new forces for the benefit of man." Carnegie was confident that this unusual formula would succeed. And he was right.

For over a century, the Carnegie Institution has sponsored creative and often high-risk science. Some of the luminaries who were supported by the Institution over the years are well known. For example, Edwin Hubble, who made the astonishing discoveries that the universe is larger than just our galaxy and that it is expanding, was a Carnegie astronomer. Barbara McClintock, who discovered the existence of transposable genes, and Alfred Hershey, who proved that DNA holds the genetic code, both won Nobel Prizes for their work as Carnegie scientists. But many other innovative Carnegie researchers who are perhaps not so well known outside their fields of work have made significant advances.

Thus, as part of its centennial celebration, the Institution enlisted the help of many individuals who have contributed to the Institution's history to chronicle the achievements of the Institution's five major departments. (Our newest department, the Department of Global Ecology, was started in 2002 and its contributions will largely lie ahead.) The result is five illustrated volumes, which describe the people and events, and the challenges and controversies, behind some of the Institution's significant accomplishments. The result is a rich and fascinating history not only of the Institution, but also of the progress of science through a remarkable period of scientific discovery.

Andrew Carnegie could not have imagined what his Institution would accomplish in the century after its founding. But I believe that he would be very proud. His idea has been validated by the scientific excellence of the exceptional men and women who have carried out his mission. Their work has placed the Institution in a unique position in the world of science, which is just what Andrew Carnegie set out to do.

RICHARD A. MESERVE
President, Carnegie Institution of Washington

PREFACE

Any life a century long deserves notice, especially when that life has been full of innovation, inspiration, and incubation of new ideas and contributions and especially when at 100, the individual is as full of life and promise as ever before. The Carnegie Institution of Washington (CIW) has had such a life and is full of such promise. As CIW President in 2001, Maxine Singer was in the position to develop ways to reflect on the past, celebrate the present, and look forward to the future. She decided to follow the traditions of the CIW itself, namely relying on research and investing in individuals with a clear set of goals and mission. One result was the lovely public volume celebrating the CIW personality, by James Trefil and Margaret Hindle Hazen (*Good Seeing. A Century of Science at the Carnegie Institution of Washington. 1902–2002* (Washington, DC: Joseph Henry Press, 2002)).

Another result is these volumes, a product of Maxine Singer's interest in capturing what is exciting about the science. To this end, she focused on existing CIW Departments and lined up single authors to write all but one of them. This one required a more complex approach, and she decided to pursue a multi-authored volume instead. She put together a team that includes leading historians of biology and the past two Directors of the Department. The group came together twice, once for a planning meeting and once after circulating drafts in order to provide comments on each of the other chapters and to work toward greater integration and communication among the chapters. That process has worked well, and the result is a rich set of original chapters looking at the CIW Department of Embryology. In addition, the editors added Jane Maienschein's introductory essay and Garland Allen's explanation of the role of the CIW Department of Genetics and its relations to the Department of Embryology, since the two were closely connected in the short-lived Division of Animal Biology. Marie Glitz has coordinated the project and has held the authors' hands through all the annoying details that make a set of chapters a whole instead of a stapled-together bunch of individual manuscripts.

This project, like any complex process, would not be possible without a team. Along with Maxine Singer, Margee Hazen got us started and provided energetic moral support through the first stages of the project. When she left the CIW, Tina McDowell adopted the project and has eased it through the

Franklin Paine Mall (first Director, 1914–18)

George L. Streeter (second Director, 1918–41)

George Corner (third Director, 1941–55)

James Ebert (fourth Director, 1956–76)

Donald Brown (fifth Director, 1976–94)

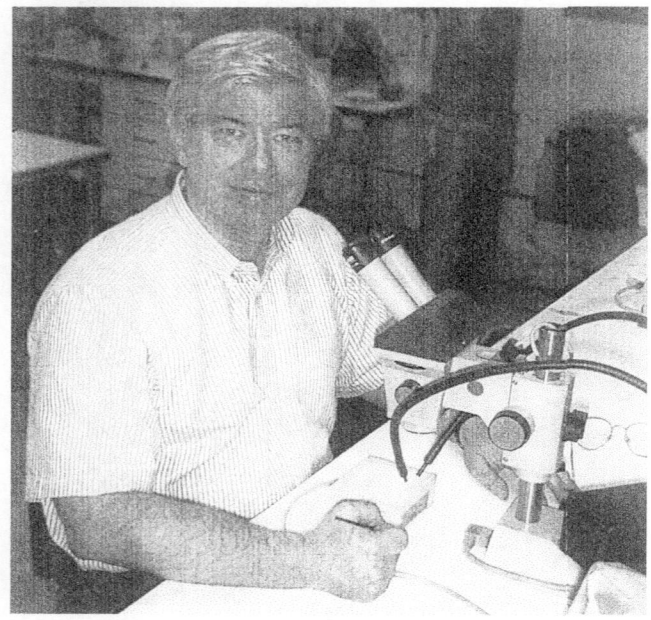

Allan C. Spradling (sixth Director, 1994–)

production process. All the CIW staff have been wonderfully supportive and helpful, for which we are grateful.

Illustrations throughout the book, unless otherwise indicated, come from the CIW Archival Collection. This collection consists of file cabinets full of photographs that complement the major archival collection of documents. The CIW Archives should serve well the interests of many historians for the next centennial project, though scientists are developing the habit of not saving documents that remain ephemeral and electronic, and the records may be better for the first hundred than for the next hundred years.

As Donald Brown noted in his final annual Department report (*Year Book* 93 (1993–4), p. 29), "A visiting committee report once said that our department functions with no trace of administration. I think that was a compliment." We all know that the Department Directors are indispensable to the CIW success, which is based on the central premise that investing in outstanding individuals will work. The Department of Embryology has succeeded in large part because of its excellent leadership by scientists who understand the possibilities of investing in exciting research, and they deserve our notice and applause. These include: Franklin Paine Mall (1914–18); George L. Streeter (1918–41); George Corner (1941–55); James Ebert (1956–76); Donald Brown (1976–94); and Allan C. Spradling (1994–).

INTRODUCTION

JANE MAIENSCHEIN

School of Life Sciences, Arizona State University

"Embryology," naturally enough, evokes images of embryos. Certainly at the beginning of the Carnegie Institution of Washington Department of Embryology in 1914, there were embryos in abundance. These were human embryos: preserved, sliced and studied in order to construct models that became the basis for human embryology textbooks and medical school training. Primate studies then provided information and understanding of embryo implantation in the mother's uterus, of material exchanges between mother and embryo, and of the entire developmental cycle through studies that would have been impossible with humans. By the second half-century of the Department's work, embryos had receded in importance. Tissue and cell cultures provided new histological information about development. Biochemistry, molecular genetics, and relations of genetics to embryogenesis took center stage. Yet, unlike other university departments, professional societies, and journals, the Carnegie Department did not rush in the second half of the twentieth century to change its name from "embryology" to "developmental biology" or "molecular biology." At heart, the research group remained concerned with the processes of development. And, yes, with embryos, through experimental embryology and then through development and genetics. Now, embryos are in vogue again, vaulted onto the front pages of local newspapers by a cloned sheep named after Dolly Parton, by stem cell research, and by the hopes for improved reproductive medicine.

This book explores the Carnegie Institution of Washington (CIW) Department of Embryology since its inception. Who did what, where, how, and why? What contribution did this department make to the development of biological understanding of embryos, and what is this group doing to lead the way into the future? In this chapter I draw especially on the annual reports from the CIW and on the papers in this volume to provide an introduction to the Carnegie philosophy and to the personality of an institution that is distributed across different places and with people who move in and out of the story. The CIW could have become an anachronism, a sort of monument to the hopes of the progressive era, frozen in time with the vision of a late nineteenth century rags-to-riches man who made good. But it has not become that. The Carnegie Institution has remained vital because of the underlying principles and the selection of good people to guide programs. The Department of Embryology has helped to keep embryos scientifically alive in the many senses that this volume discusses.

Andrew Carnegie and his Institution

The story begins with Andrew Carnegie, and indirectly with Carnegie's mother. His mother's dominance in his life undoubtedly shaped Carnegie's own drive to succeed and to concentrate on business and community, since she kept him from marrying and developing strong independent ties during her lifetime. Considered a "robber baron" by his critics or a "captain of industry" by his supporters, Carnegie made money in steel – a lot of money. When he succeeded beyond even his imagination, he resolved to put that money to good use. His philosophy of "scientific philanthropy" called for not just scattering funds to individual isolated causes or leaving large sums to one's heirs but rather for investing in the future. Outlined in his "Gospel of Wealth," Carnegie's ideas rested on the assumption that it was better to educate and support than to give handouts on which recipients might become dependent. Wealth must be properly administered, he insisted, and "It were better for mankind that the millions of the rich were thrown into the sea than so spent as to encourage the slothful, the drunken, the unworthy."[1] He gave library buildings, but left it to the community to provide the books and the librarians. He gave to universities, particularly to the Tuskegee Institute, Hampton College, and Berea College rather than to better established schools and made sure that the programming was top quality and the money well invested. And he established the independent Carnegie Institution of Washington to promote scientific research (Fig. 1.1).

In 1901, Carnegie concluded that it was time for an "institution of higher learning" in Washington. Yet he decided against establishing a university there that would compete with other universities. Instead, he settled on an independent research organization. The lovely centennial volume by James Trefil and Margaret Hindle Hazen, entitled *Good Seeing. A Century of Science at the Carnegie Institution of Washington, 1902–2002*, outlines those early discussions and the history of the institution overall. Clearly, Carnegie was inspired by John D. Rockefeller's new medical research institute in New York. His enthusiasm for supporting the individual "genius" pointed to an institution that would allow those individuals to try new ideas in a climate unfettered by the needs to teach or to sell ideas to industry. His goal was to promote both basic research, with "investigation, research, and discovery 'in the broadest

Figure 1.1 Andrew Carnegie "America's Most Eminent Business Man."

and most liberal manner,'" and application, fostering "the application of knowledge to the improvement of mankind."[2] Given Rockefeller's emphasis on medical research, Carnegie resolved to look in other directions and not to include clinical medicine.

The new Carnegie Institution began in 1902, with Daniel Coit Gilman as President. Gilman had served as first president of the University of California from 1872 until he moved to become first president of the newly founded Johns Hopkins University before then accepting the new challenge of heading the CIW and developing its mission. At first, the institution awarded individual grants. In the biological sciences, some of the most visible funding went to the individualistic Luther Burbank, and some of the most important early support went to George Harrison Shull at the Cold Spring Harbor Laboratory. Burbank was the sort of "genius" Carnegie sought to invest in, but was idiosyncratic and unable to share his individualist approach with others. The Carnegie sent Shull to study with Burbank to learn his scientific methods, but Shull concluded that they were actually not scientific at all and perhaps not very methodical. The resulting "Burbank problem," where Carnegie favored Burbank while the trustees were more skeptical about what Burbank actually offered for the longer term, clearly influenced the selection process and the organizational structure for further Carnegie awards.[3]

The Carnegie Institution opted for a combination of individual awards to selected geniuses for short-term support and with the apparent expectation that there would be results in the form of reports and publications. Carnegie had written that "You know my own opinion is that no big institutions should be erected anywhere." Instead, "exceptional men should be encouraged to do their exceptional work in their own environment." Carnegie had concluded that "There is nothing so deadening as gathering together a staff in an institution. Dry rot begins and routine kills original work."[4]

Yet this did not mean that the Institution had no place. In 1909, the trustees dedicated an administration building at 16th and P Street in Washington. In addition, various research laboratory sites have come and gone over the century, as appropriate for the work at hand and often in partnership with other institutions and individuals. Genetics found a home at Cold Spring Harbor Laboratory, on Long Island, and in other places like Thomas Hunt Morgan's laboratory at Columbia University. Embryology centered in a sequence of at first borrowed, and then specially-designed, laboratory buildings associated with the Johns Hopkins University.

The Department of Embryology

In 1913 Franklin Paine Mall applied for Carnegie support for his work on human embryos. As Nick Hopwood has documented in an outstanding study, *Embryos in Wax*, close examination of human embryos had gained considerable attention in the preceding decades, notably through the work of Wilhelm His and Franz Keibel.[5] These researchers sought through detailed anatomical and histological studies to trace the changes in structure from the very beginning of embryonic life. That is, rather than just assuming that life really begins at the traditional forty days or at the point when germ layers are well defined as many morphologists had assumed, these embryologists believed that it was at least important to assess the significance of the earliest stages. Presumably, the importance of structure does not begin all of a sudden at a later point, but exists from the beginning. At the very least, we should know more about the entire embryological sequence. To that end, they collected, described, and modeled as many stages of the developmental process as they could find, though initially these necessarily focused on later stages since those were the ones most easily available. Embryos in the earliest stages are nearly invisible, and it took more experience to know even what to look for or to know what the tiny embryonic thing was once it was observed.

Following other leading American anatomists, Mall went to Leipzig to study with anatomist/embryologist Wilhelm His and in his role as anatomist at the Johns Hopkins Medical School began to amass his own collection of

Figure 1.2 The "S. S. Franklin P. Mall," named after the first Director of the Department of Embryology.

human embryos. There is no better way to learn than by doing, he argued, and no better way to teach than with observations of models and specimens to inform study of the static textbooks. On February 20, 1913, Mall received Carnegie Grant No. 874 for $15,000. Work began right away to catalog the existing collections and to secure the collections and records in fireproof facilities. As Mall put it, "A vigorous campaign has been carried on for new specimens of human embryos," reaching half the physicians in the USA and many internationally.[6] This aggressive strategy paid off with new specimens and increased visibility for the collection, presumably helped by the stability afforded by a substantial grant and institutional support from the Carnegie Institution generally (Fig. 1.2).

Mall was made director of a new Department of Embryology, a position he held until his death in 1917. By 1916, Mall was reporting in the annual *Year Book* that while it had taken ten years to get his first 100 embryos, five years for the next 100, three years for the next, and two years for the next, 400 specimens per year had been pouring in since Carnegie support had begun in 1913. He noted that over 500 persons had contributed to the collection (apparently not counting all the mothers who were obviously but in many cases obliviously involved).[7] With Carnegie visibility and authority, the project attracted support from the medical profession generally and even from the State Board of Public Health of Maryland, which instructed physicians in the state to send their specimens to the collection for the purpose of advancing our collective knowledge.

Figure 1.3 Carnegie Laboratory of Embryology, Modeling Department, 1921.

In supporting the project, the Carnegie Institution soon officially opened the Department of Embryology on the Johns Hopkins Medical School campus, with Mall as Director. Within a few years, they were expanding the facilities, adding square footage and facilities for photography, machinery to support making the models, and expanded storage for the collection and the records. By 1915, Mall had formally transferred ownership of his collection of over 2000 specimens to the Carnegie.

Over the next decades, researchers sectioned the specimens, recorded the sections with photographs and drawings, and preserved the materials themselves in fireproof vaults with considerable attention to the acknowledged irreplacability of the collection. In many cases, to augment the specimens themselves and the detailed records about their collection and their analysis, the researchers had models constructed. As His had in Germany to develop his collection, Mall's group hired sculptors to ensure quality and accuracy (Fig. 1.3). By 1914, Mall had hired His's former student and collaborator Franz Keibel from Germany. Keibel had considerable experience in preparing the embryos, so this was a major advance that moved what had initially been a collection of embryos to a major and long-term project of considerable embryological and medical significance. Mall also attracted George Streeter from the University of Michigan, whose work focused on development of the nervous system. And cytologists such as research associate Edmund Cowdry assisted with histological studies while Warren and Margaret Lewis contributed other cell studies.

The result was an impressive group of researchers, established with Carnegie funding and cooperation with Johns Hopkins, at a time when German hegemony in the fields of anatomy and embryology was being considerably undercut by the onset of the First World War. This period of research led to the set of what the group codified as twenty-three distinct stages from fertilization to the eight-week, or fetal, stage. The Carnegie stages, solidified by Streeter, became the standard worldwide for human embryos, and the staff provided a public service for physicians by comparing with the normal stages the abnormal, spontaneously aborted specimens acquired from autopsies sent in by physicians.

Streeter described in the 1918 *Year Book* report the research that Mall had been pursuing at the time of his death in November 1917, including calculations that for every twenty spontaneous abortions, there are eighty full-term births; and that an additional thirty "monsters" are born to every 5,000 pregnancies. In addition, the Carnegie group had made further progress in detailing the timing and sequence of steps in human fertilization and embryo implantation. Streeter was enlisted to serve as Acting Director of the Department for one year after Mall died, and then served as Director until he retired in 1940. As with all the other departments, Carnegie researchers do not receive tenure, and many leave after establishing a research record in this

rich and supportive environment. Fortunately, a few of the leading scientists
have remained and have taken on important leadership and administrative
positions and stayed with the Carnegie throughout their careers.

By the next year, after his first full year as director, Streeter had had time to
reflect on the directions of the Department. He reported that they remained
focused on human embryology as their primary problem, including microscopic study of cell structure and gross anatomy of organ systems to understand the body as a whole. They were discovering the value of comparing
not only the standard normal, but also pathological specimens to appreciate the factors involved in producing abnormalities. This was obviously of
medical importance though not involving clinical research directly and once
again reflects the practical aspects of the Carnegie mission. Already there
were plans for a new building to provide more space. Warren Lewis had
been made a research associate to the Department and had, with his wife
Margaret, developed valuable tissue culturing techniques that had already
proved innovative for culturing embryonic tissues and expanding cytological studies. Under Streeter, the embryological work continued, but Streeter's
own contributions soon brought that line of research to a natural end that
pointed in new directions.

By 1973, the emphasis of the Department had changed so much that the
collections were really no longer used. They were moved to the University
of California at Davis, and then again in 1990 to the National Museum
of Health and Medicine of the Armed Forces Institute of Pathology, where
they reside today. The collection has recently been digitized and is available
through the internet as a resource for the medical and research community
and for historians.[8] Adrianne Noe discusses this phase in the history of the
Department both in this volume and in her other work cited there.

Primate and comparative studies

Following the emphasis on anatomy with the human embryos came comparative studies with other animals such as chicks, pigs, cows, and then
primates, with a focus on physiology. Elizabeth Hanson's chapter in this volume, chapter 3, describes and explains the importance of the primate colony
for the CIW study of embryology. It was during Streeter's chairmanship of
the Department that the monkeys arrived. One of Mall's students, George
Corner, studied anatomy and had become particularly interested in the cycle
of reproduction in mammals. He began his studies of rhesus monkeys in a
laboratory at the Johns Hopkins, and continued that work through the CIW.
He then moved to head the Department of Anatomy at the University of
Rochester Medical School from 1923 to 1940.

The initial small group of monkeys became a large colony of rhesus
macaques, and the Carnegie researchers' continuing studies achieved such
success that the Department recruited Corner to return and follow Streeter

as Director of the Department from 1941 to 1955. Corner's role as Director indicates the direction of research during this period. As Hanson shows in chapter 3, Corner's rhesus monkey colony made possible detailed study of this mammal thought to be closely related to humans and with a menstrual cycle like that in humans. The studies played an important role in focusing serious biological attention on reproductive biology. Adele Clarke's chapter in this volume, chapter 4, demonstrates the nature and importance of that reproductive study, which shaped and even substantially helped to create a disciplinary field of study. Hanson shows that the decision to establish and sustain such a monkey colony required considerable continued investment. As Clarke demonstrates, that investment paid off well in both basic and practical knowledge, in this case to benefit women as part of the Carnegie objective of seeking "improvement of mankind." Although the primate colony was eventually transferred, as Hanson explains in chapter 3, the reports of the 1930s and 1940s are full of discoveries about endocrinology, physiology, and neurology (related to primate) and gynecology (related to human).

Streeter noted in his report for 1936 that there were differences of opinion about just how far the researchers ought to be pressed to develop the medical applications of their work, and about how to organize that work. "The question is raised as to how much freedom should be given to the independent investigator." Should there be dedicated institutes just for the study of cancer, for example? This approach would be too regimented, Streeter concluded. He noted that in pursuing other studies the Department of Embryology had made important discoveries about the nature of tumor development, for example, and the Department of Genetics had added knowledge about tumor heredity even though cancer research was not their primary mission. This argued against single-mission medical laboratories and called for the importance of supporting research into "the fundamental facts upon which an understanding of the nature of cancer must eventually rest," or a call for basic research.

Furthermore, different groups, working quite independently of each other, were making discoveries that complemented each other and added up to significantly advanced knowledge. Therefore, "It is obvious that intercommunication between the groups should be frequent and full, in so far as this can be brought about without infringement upon the backgrounds and approaches of the respective groups. Such an intergroup awareness is facilitated by our administrative organization as a division."[9] Distances between the individual labs, such as Embryology at Johns Hopkins and Genetics at Cold Spring Harbor, should not be allowed to become a barrier to exchange of ideas and free and open cooperation. Any university today would be happy with that emphasis on collaboration, which is something they all seek – or at least say they do.

The CIW sought to realize those hopes by coordinating the Department of Embryology and the Department of Genetics, plus the Nutrition Laboratory in Boston and the Tortugas Marine Laboratory in the Florida Keys, into a Division of Animal Biology starting in 1934. Streeter served as Chairman and explained that the separate biological studies had been "in each case located where it seemed they could be best conducted." The tendency to overlap and to relate to one another had become sufficiently strong, however, that by 1934 formal coordination had seemed desirable.[10] Yet, as Garland Allen explains in chapter 6, this volume, this spirit of cooperation at times remained more rhetorical than real across the areas of embryology and genetics. By 1941, reports once again came from the separate departments.

Recording cell and tissue development

Hannah Landecker explains the contributions of Warren Harmon and Margaret Reed Lewis in chapter 5, this volume. Rather than theoretical originality, they brought technical skills to the study of cells and tissues. In particular, the techniques to culture tissue and cells outside the body afforded the opportunity to record what happens in the culture. Ross Harrison had developed the very first tissue culture techniques, using hanging drops to culture nerve fibers and demonstrate that they experience protoplasmic outgrowth that appears to be just like that in normal development.[11] Harrison first pursued this work at Johns Hopkins, before moving to Yale, and he worked with the Lewises. While Harrison gave up the technique as not central to the problems he wished to pursue, the Lewises carried development of the technique further. Landecker's account of their work focuses on the intriguing decisions to record the steps of development on film.

Clearly, embryonic development is a process, and it takes place through time. The fascination with capturing the movement is obvious. The idea that following cells and tissue changes during every step of the process rather than just at defined "stages" must have been compelling. Furthermore, the attraction is enhanced by the possibilities for speeding up and slowing down the film to observe details even better. As Landecker explains, the Lewises contributed to a significant shift in anatomical and embryological studies, toward seeing the cell as a dynamic contributor rather than passive respondent in the developmental process. This work began while Streeter was Director, during the 1920s, and continued under Corner's direction into the 1940s.

What the Lewises contributed, as Landecker shows so nicely, was techniques. They helped to develop infrastructure that allowed the research to succeed. The embryo collection, the primate colony, and the tissue culturing and video recording all provided considerable support for investigations by others, both within and beyond the CIW. The Lewises therefore provide a beautiful example of the wisdom of the Carnegie philosophy. Investing in people and supporting their innovations and encouraging them to work together produced a lively intellectual community. The case of the Lewises makes clear that within this context, not everyone has had to be a leader in advancing theory, yet they could still be central players in a team built on the healthy cooperation, collaboration, and mutual respect that still characterizes the CIW Department of Embryology philosophy.

Embryology at 25 years

On April 4, 1939, in the Department's twenty-fifth year and his last as Director, George Streeter reflected on their contributions. They were receiving about 600 embryos a year, he noted, and providing a service to the physicians who had sent them through the accumulated collation of embryological knowledge and detailed descriptions of developmental stages. Therefore, retaining strong connections with the medical community, facilitated by the Johns Hopkins location, was essential. In addition, the researchers had been busy. In the first quarter century, he counted 1,148 articles by staff members and associates, plus twenty-seven monographs in the Carnegie *Contributions to Embryology* series.

The research had not always gone as expected, Streeter noted, for "Apparently the progress of research can be predicted only to a limited degree and we must recognize that opportunism plays a large part in discovery." Indeed, "advantageous alterations in course" are important, and "it is well to be aware of the limitations of rigidly planned and far-flung research."[12] He acknowledged that the Department had taken up work and moved in directions not originally foreseen. Yet there nonetheless remained a driving goal by the Department of Embryology underlying all the changes, namely to study embryonic structure and how it develops. This included the structure and function of component parts, and the factors that shape them throughout a lifetime. Yet within this broad mandate, it was wisest to invest in the best people and let them do their work. That is the recurring message from the CIW leadership.

By 1939, there was already tremendous interest in genetics and the mechanisms of heredity. Many felt that the mechanisms of reproduction would be found in genetics. Yet Streeter urged that "there still exists a large gap between where the geneticist leaves off and where the embryologist begins." We do not know the forces that direct embryogenesis, nor how genes affect or effect their control, if they in fact do. Geneticists have techniques for demonstrating correlations but not yet for establishing causes. Embryologists do have techniques for following how one stage gives way to the next, or the appearance at least of causal connection, and Streeter's message was that it was incumbent upon embryologists to continue their work. Biochemistry, endocrinology, genetics, study of growth in cancers: all are important and part of understanding development. There are "stimulating advantages of such interchange of ideas and cooperative investigation. There can be no doubt of the great value of the combination of genetics and embryology in the important undertaking which is now occupying so large a share of our thought."[13] Let us not lose sight of the embryos, Streeter seemed to be saying.

He also noted the research on primates that had begun in the 1920s. With humans we can only work with those embryos that happen to become available when physicians bring them in. Yet there is great advantage in understanding the full reproductive cycle in humans as well as other animals. Researchers associated with the Department had studied how an egg fastens to the uterine lining and the early stages of placental formation. They had already made considerable progress in understanding exchanges between mother and developing embryo.

In addition, with animals including primates, researchers could carry out tissue cultures and cell cultures. They could extract cells and culture them, thereby allowing research on living materials rather than just the dead and dissected materials afforded by the preserved specimens. Observing many specimens in animals allowed them to interpret what they saw in the few human cells and embryos to which they had access. Warren Lewis's development of a system to capture the observation in film using the relatively new techniques of making motion pictures led to increased data and understanding of the character and details of cell movements.

Streeter pointed to possible future research directions, including the prospects for using radioactive tracers to track the movement of materials from mother to embryo. Yet he also warned that they must not become like Aesop's dog, which in the excitement of seeing its reflected image with bone in the water, dropped the bone it already had. By implication, the CIW Department should not drop important ongoing work to chase trends and fads. They should therefore retain their solid foundation in first-rate scientific exploration while pursuing innovations where and only where they made sense. Above all, what CIW could offer this research was the capacity to work across disciplinary boundaries, the flexibility to promote cooperations, as between geneticist and embryologist, and to incorporate techniques and tools from a wide diversity of fields, and thus the ideal climate for incubation of innovations. Streeter retired with an optimistic outlook on the CIW efforts.

The Department's wartime effort

In his 1944–5 report, Director George Corner noted that in the fourth year since war had begun, he had to offer the smallest review of research since the Department had been established. His staff had been diminished, and had been distracted by "emergency duties" and by "the general disturbance." Nonetheless, Corner reported on new microtomic techniques used by the modeling and technical team, the continuation of the monkey colony, and participation in cooperative efforts stimulated by the war effort. Indeed, he cited "Embryology as a cooperative science" in the title of his report.

This becomes a theme: cooperation and teamwork, for "Among the gains brought by this way, in partial compensation for its destruction and misery, surely not the least is this cross-fertilization of the various sciences, which results not only in immediate practical advantages, but also in new thinking about fundamentals. The synthesis of ideas thus achieved is not lost even if in times of peace the pendulum swings again necessarily toward individualistic research." This was a notable change of tone for the CIW, emphasizing the interactions more than the individual research efforts. Reports of the next years continued to stress the prospects for interconnections and cross-fertilizations as a justification for the various separate lines of research. "At any moment," Corner noted, the results of "pure" science may become useful. Indeed, "The understanding of man's place in the animal world," such as those promoted by the embryological group, "necessarily influences the whole structure of human education, lawmaking, and philosophy."[14]

By a decade later, Corner was reporting a huge influx of researchers from around the world. The recitation of publications, description of research, and list of researchers was impressive, indeed, and showed how far the Department had come in these glorious post-War years. Usefulness was measured far more clearly in terms of value to the understanding of embryology and fundamental biological problems than to external applications.

Carnegie at 50

Vannevar Bush was president of the CIW during its fiftieth year, in 1951–2. The event promoted reflection. The Institution continued to proceed in just the way the founder had envisioned, promoting mutually beneficial basic and applied research. The focus also remained on the individual scholars, with investment in the talented expected to yield results. Bush believed strongly in such investment, with the primary responsibility to invest in fundamental research without regard to the potential payoff or industrial application that would follow if and when appropriate. Much remained the same, therefore.

Yet the departments were continually evolving, and that was good. After all, though the scientists might each differ "in what they are trying to accomplish," they conducted scientific research the same way. Piece by piece, the selected individual researchers have added to our pool of knowledge that is available for public good. The CIW programs have begun with basic research, publishing openly, with laboratories open to any serious-minded visitors and open to opportunities for developing patents should those arise. Perhaps the Institution could and should help to hasten the time from research to development, Bush reflected, since the bridge between basic and applied research is often ineffective with too many hurdles. In particular, he pointed to George Shull's pioneering work on hybrid corn at Cold Spring Harbor and the time it took before the idea made it to market. Perhaps the Institution should have promoted the development more directly, but it was not clear how. Other devices to help with childbirth and information for textbook training of physicians provided more evident moves from the lab to the practical world.

Notably, when Bush turned to the contributions from the Department of Embryology for that year, it was already organs rather than embryos that received top billing. Discoveries of neural crest cell formation shed light on development of the eye and the neurology of vision. Kidney formation in humans provided another example. "In this respect, as in many other details of embryonic development," Bush noted, "our species is more like other mammals than has sometimes been thought." X-ray studies, and in particular X-ray motion pictures were revealing details about umbilical arteries, and combined with studies of the chemistry of uterine muscle contractions were illuminating many details of the reproductive process. CIW researchers were "getting very close to the fundamental problems of the specific manner in which hormones exert their extraordinary physiological effects upon their target organs."[15]

This might be "fundamental research," but Bush and the CIW research staff remained very aware of the potential benefits for greater understanding of important biological processes and also for potential medical applications. The fundamental or basic science easily merged with the applied, Bush emphasized, even in detailed and specific studies. Together the individual studies asking different questions with different techniques combined, through communication and cooperation, to advance knowledge for the "improvement of mankind."

In his last report as Director in 1954–5, Corner had reflected on the first forty-two years of the Department, outlining a formidable portfolio of diverse studies. As Donald Brown discusses in chapter 7, this volume, Corner had raised questions about future directions for the Department and how best to remain innovative and adaptive to changing environments. The decisions made reflect the impact of focusing on techniques and infrastructure, and the value of investing in people and allowing them enough time and support to incubate innovations without demand for immediate results.

The selection of Mall, Streeter, and Corner as the first three Department Directors reflects the strong medical connections of the Embryology Department. Embryology was a practical and medical science, and the CIW researchers saw it that way. They brought to bear studies of cells, tissues, and other basic biological tools and questions. Yet the emphasis remained medical and the ultimate goals practical through this time. That changed with the next director and after the mid-1950s and the advent of DNA studies.

The second half-century of the Department of Embryology has therefore looked quite different from the first, demonstrating the wisdom of retaining flexibility and mobility in investment that Carnegie and the early trustees had emphasized from the beginning.

Genetics and evolution

In this volume and elsewhere, Garland Allen has discussed the CIW's investment in genetics and experimental evolution. Most notably, this occurred through individual grants to fund Thomas Hunt Morgan's fly room at Columbia and for Nettie Stevens' studies of chromosomes at Bryn Mawr, and through institutional funding for the Department of Genetics at Cold Spring Harbor. For Cold Spring Harbor, Charles Davenport had a vision of a Station for Experimental Evolution. He saw the importance of studying heredity, since inherited variations are the raw material for the evolutionary process. Clearly, he concluded, traits run in families and therefore are inherited. Therefore, Davenport suggested that to promote intelligent evolution, we should begin with data about which families carry desirable and which undesired traits. Eugenics made perfect sense to Davenport, and was consistent with commitment to advancement of the general public health. Surely a visionary like Andrew Carnegie and his Institution's trustees would be committed to this interpretation of the "improvement of mankind."

So they were. As Allen explains in chapter 6, this volume, Davenport approached the Institution for funding in 1902. At first, the CIW funded Davenport through the Station for Experimental Evolution at the Cold Spring Harbor Laboratory, begun in 1904. Then in 1918, that station was combined with Davenport's separately funded Eugenics Record Office into the Carnegie Institution Department of Genetics. Though the Department continued until 1962, after 1940 the Eugenics Record Office was closed and the emphasis placed on the genetics research of a few individual investigators, including two of the most important genetic researchers of the twentieth century, Alfred Hershey and Barbara McClintock. In 1962, Carnegie president Caryl Haskins in effect gave the Department to the Cold Spring Harbor Laboratory, with the understanding that the CIW would continue to fund those two researchers. From 1962 to 1971, what was called the Genetics Research Unit continued, headed by Hershey, who along with his predecessor Milislav Demerec had considerably raised the quality of research pursued.

Carnegie in the post war era of expanding science

The 1950s brought changes, of course, with the rise of the National Science Foundation, increased funding for the National Institutes of Health, democratization of the universities as soldiers continued to use their GI Bill benefits to pursue education and a promised better future. It was a period of adjustment, with the CIW guided by Vannevar Bush. In 1956, Caryl Haskins became President at the same time that James Ebert became Director of the Embryology Department. By 1961, with the sixtieth anniversary of the Institution, the changes had become more visible.

In his presidential report in the 1961–2 *Year Book*, Haskins grew reflective. He cited the early decisions to make the CIW largely an operating rather than granting institution, with localized smaller laboratories around the separate departments but with flexibility as well. This had served the Institution well, for each unit remained relatively independent and "able to seize the initiative in new and appropriate fields as they appear, yet all sufficiently connected so that they may be of mutual assistance as the needs arise." Organization in itself is not important, except in that it gives a framework for the research.

The philosophy of an institution matters, and the CIW philosophy was to promote the creative individual, assuming that the result would be incubation of original thinking and innovative science.

> And with this goes the philosophy, equally deep-seated and equally important, that this freedom from fixed commitment applies to fields of endeavor as well as to men: that high mobility within specific fields, that of unfettered crossing of fields, that the fashioning of unconventionally wide-ranging programs, are subject only to the limitations imposed by Nature and by the judgment of the gifted and discriminating investigators, and that making this mobility and this flexibility possible is the principal objective of the Institution.

Again, any twenty-first century research institution would be pleased to point to a history of successful promotion of such creativity and would love to know how they could achieve the successes that the CIW experienced.

Haskins realized that the world had changed since Andrew Carnegie had first laid out his vision. By 1961, science had seen tremendous growth and was obviously going to continue its extraordinary expansion of personnel and intellectual development. Could it possibly be that the reliance on small and mobile groups of individual researchers, encouraged to be creative and innovative, was no longer the way to promote scientific discovery? "Is it possible that we are witness to a profound revolution in the very character of research itself," and must we develop larger and differently organized teams? This question about organization "touches on the nature of scientific truth itself," Haskins thought, and it touches on the CIW faith in the "distinguished, unfettered individual." Was this the end of an era after "only" sixty years?

No, of course not. Yes, there was room for other ways of organizing science as well. But the most recent research in genetics, for example, showed that it was "abundantly clear that the essential qualities and requirements of inquiry at the very frontiers of man's knowledge of his universe do not now, and in all probability will not in the foreseeable future, differ significantly from those of our classical scientific past." Surely scientific inquiry would continue as it had, and the Institution should continue its wise investment in people, providing them whatever material and human support might be required for the production of knowledge. There is a heroic quality in the expressions of faith. And an inspiring conviction that the CIW leaders carried a "heavy responsibility of the keeper of a vision." It was their task to facilitate creativity, remove obstacles, and encourage the "priceless jewels" that individual innovators represented to continue their work. Quoting Chaucer, Haskins reminded his readers that it made sense to continue the CIW traditions rather than rejecting what had proved to be a sound philosophy, for "Out of the old fields cometh the new corn."[16]

Meanwhile, James Ebert was pointing the Department of Embryology in new directions. Not at all giving up traditional problems of embryology, he saw the Department as embracing new techniques of genetics and molecular biology to attack those problems. Even before they began their terms, by 1955, Ebert and Haskins had apparently agreed to phase out the human and primate embryological work.[17] Instead, the Embryology Department would explore the most fundamental embryological problems of differentiation, growth, and morphogenesis with new techniques. Clonal cell lines would provide material for exploring genetic transduction. Biochemistry, physical chemistry, and optical methods would allow close examination of the fine structure of cells.

Whereas Mall and Streeter had emphasized morphology and anatomy, Corner had brought physiology, biochemistry, and biophysics to embryology. In turn, Ebert brought genetics to development and helped transform embryology at CIW. Rather than correlations, population studies, or the role of transposable elements pursued by the Department of Genetics and particularly by Barbara McClintock at Cold Spring Harbor, the Embryology Department would focus on "genetics by gene isolation" as researcher and later Department Director Donald Brown put it. Gene amplification, isolated purified genes, and RNA all provided material for the study of the role of genes in development. Brown's chapter in this volume discusses the period that began after Ebert arrived and that so notably redirected the Department. Brown provides further detail and insight into the biological inquiry of this important transformational era for the Department of Embryology and for embryology itself. He also emphasizes that science remained the focus of the Department and the Directors (Fig. 1.4).

In addition, as science became more complex and called for multidisciplinary approaches, the CIW cultivated an emphasis on innovation along

Figure 1.4 Donald Brown, Department of Embryology's fourth Director, at the bench.

with the attitude of cooperation and collaboration. As current Department Director Allan Spradling says in his chapter in this volume, chapter 8, the current Department research team is proud to say "we do not know." He notes that researchers are expected to generate first-rate and ground breaking science, not necessarily more publications or more grants just for the sake of having more. Follow the research and innovations, and be patient in waiting for results if necessary. Spradling gives us insights into the current opportunities and limitations for a Department of Embryology in this era of translational genomics and developmental genetics. As Spradling put it so perfectly in a recent annual report, "Genomic research is frequently viewed by the public, and even in some scientific quarters, as a relatively new development. In reality, though, this institution and the Department of Embryology have been striving to decipher gene structure and function for most of the last 100 years"[18] (Fig. 1.5). In different ways at different times, researchers have continued to work toward methods and approaches for answering the

Figure 1.5 Members of the Allan Spradling Laboratory team introducing purified genes into fruit fly embryos, 1983, including Terry Orr-Weaver (postdoc, standing), Suki Parks (graduate student, seated), and Joe Levine (technician).

fundamental questions about embryonic development. For 100 years, the CIW Department of Embryology has been a leader in cooperation, innovation, and incubation of new ideas in a changing world.

Notes

1. Andrew Carnegie, "A Gospel of Wealth," cited in James Trefil and Margaret Hindle Hazen, *Good Seeing. A Century of Science at the Carnegie Institution of Washington. 1902–2002* (Washington, DC: Joseph Henry Press, 2002), p. 16.
2. Trefil and Hazen, especially pp. 21–35.
3. Trefil and Hazen, pp. 31–33.
4. Trefil and Hazen, p. 33.
5. Nick Hopwood, *Embryos in Wax. Models from the Ziegler Studio* (Whipple Museum of the History of Science, University of Cambridge, 2002).
6. Franklin Paine Mall, Carnegie Institution of Washington *Year Book* 13 (1913–14), p. 290.
7. Franklin Paine Mall, Carnegie Institution of Washington *Year Book* 16 (1916–17) p. 109.
8. http://nmhm.washingtondc.museum/collections/hdac/Education_Projects.htm for the Visible Embryo Project.
9. George Streeter, Carnegie Institution of Washington *Year Book* 37 (1936–7), pp. 3, 4.
10. George Streeter, Carnegie Institution of Washington *Year Book* 34 (1934–5), p. 3.

11. Jane Maienschein, *Transforming Traditions in American Biology, 1880–1915* (Baltimore: Johns Hopkins University Press, 1991), especially chapter 9.
12. George Streeter, "Carnegie Institution of Washington. Memorandum on Department of Embryology," 4 April 1939, CIW Archives, memo, pp. 3, 4.
13. Streeter, 1939 memo, p. 7.
14. George Corner, Carnegie Institution of Washington *Year Book* 44 (1944–5), pp. 90, 91, 93.
15. Vannevar Bush, Carnegie Institution of Washington *Year Book* 51 (1951–2), pp. 16–17.
16. Caryl Haskins, Carnegie Institution of Washington *Year Book* 61 (1961–2), pp. 5, 6, 16, 25.
17. Philip Abelson, Carnegie Institution of Washington *Year Book* (1976–7), 75th year, p. 33.
18. Allan Spradling, Carnegie Institution of Washington *Year Book* 99/00 (1999–2000), p. 43.

Bibliography

Abelson, Philip, Carnegie Institution of Washington *Year Book* (1976–7), 75th year, p. 33.
Bush, Vannevar, Carnegie Institution of Washington *Year Book* 51 (1951–2), pp. 16–17.
Carnegie, Andrew, "A gospel of wealth," in James Trefil and Margaret Hindle Hazen, *Good Seeing. A Century of Science at the Carnegie Institution of Washington. 1902–2002* (Washington, DC: Joseph Henry Press, 2002), p. 16.
Corner, George, Carnegie Institution of Washington *Year Book* 44 (1944–5), pp. 90, 91, 93.
Haskins, Caryl, Carnegie Institution of Washington *Year Book* 61 (1961–2), pp. 5, 6, 16, 25.
Hopwood, Nick, *Embryos in Wax. Models from the Ziegler Studio* (Whipple Museum of the History of Science, University of Cambridge, 2002).
Maienschein, Jane, *Transforming Traditions in American Biology, 1880–1915* (Baltimore: Johns Hopkins University Press, 1991).
Mall, Franklin Paine, Carnegie Institution of Washington *Year Book* (1913–14), p. 290.
Mall, Franklin Paine, Carnegie Institution of Washington *Year Book* (1916–17) p. 109.
Spradling, Allan, Carnegie Institution of Washington *Year Book* 99/00 (1999–2000), p. 43.
Streeter, George, Carnegie Institution of Washington *Year Book* 34 (1934–5), p. 3.
Streeter, George, Carnegie Institution of Washington *Year Book* 37 (1936–7), pp. 3, 4.
Streeter, George, "Carnegie Institution of Washington. Memorandum on Department of Embryology," 4 April 1939, CIW Archives, memo, pp. 3, 4.
Trefil, James and Margaret Hindle Hazen, *Good Seeing. A Century of Science at the Carnegie Institution of Washington*, 1902–2002 (Washington, DC: Joseph Henry Press, 2002), p. 16.
Visible Human Embryo Project, http://nmhm.washingtondc.museum/collections/hdac/Education_Projects.htm.

2

THE HUMAN EMBRYO COLLECTION

ADRIANNE NOE

*National Museum of Health and Medicine, Armed Forces
Institute of Pathology, Washington, DC*

Introduction

The hunt was on. When Yale physician Elizabeth Maplesden Ramsey described "her" prize specimen for the Department of Embryology of the Carnegie Institution of Washington (CIW) in 1938, she joined throngs of others who had participated in the greatest effort to collect, organize, and study human embryos. The specimen she described was among the youngest embryos ever accessioned into the collection (thirteen to fourteen days old); nevertheless, she pointed out the importance of work that had gone before by acknowledging other extremely young embryos and the contributors for whom they were named – the so-called Peters and Miller embryos, estimated to have been a day or so younger. And the CIW-funded collecting efforts of Boston physicians Arthur Hertig and John Rock were bearing ever-younger materials. On the experience of acquiring and studying younger and younger specimens for the collection, she wrote:

> By the shores of Charles, in Boston,
> By the shining Boston Harbor,
> Stands the lab of Dr. Hertig,
> Famous Eggman, Arthur Hertig.
> Close Beside the Free for Women,
> Close to Massachusetts General,
> Flooded with a Crimson radiance.
> There, this learned famous Eggman
> Nursed a little Harvard Ovum,
> Rocked him in a bath of Bouin's
> Wrapped him soft in gauze and cotton,
> Stilled his fretful wail by saying,
> "Hush, the Yale bulldog will hear thee."
> Many things the wise man taught him

Of the eggs that shine in legend.
Showed himTeacher-Bryce and Miller,
Peters crowned with early vilii . . .
Warning said the sage-old Hertig,
"Go not forth, O Harvard Ovum!
To the realm of Dr. Streeter,
To the proving place of ova
Lest his microtome should pain you,
Mr. Miller maybe stain you,
Embryologists gaze upon you
And the fierce Yale Bulldog lick you!"
But the fearless Harvard Ovum
Heeded not the wise one's warning.
Forth he strode right to Carnegie,
Crying who's as young as I am?
Who, like me, is only 12 days?
Throwing down his gage of battle,
Reckless, challenging Yale's glory.
Who shall say what thoughts and visions
Fill the fiery brains of young eggs?
Who shall say what dreams of conquest
Filled the heart of Harvard Ovum?

Ramsey penned this while *en route* to Baltimore from New Haven, and stopped only when she had "no further time to desecrate Longfellow."[1] In these few lines, she identified many of the themes and individuals important in any historical consideration of the CIW Department of Embryology: the "race" to contribute embryos of ever-younger age, the technological processes used to study the materials, the characterizations of morphology-based staging, and the names of some of the most prominent persons in the enterprise. Likewise, these are the themes of this chapter on the development and use of the Department's collection.

Efforts to assemble human and animal embryos were not new with the Department, with its founder, Franklin Paine Mall (Fig. 2.1), or even with his mentor, the embryologist Wilhelm His. One of the most influential tomes of its day, His's 1874 *Unsere Körperform und das physiologische Problem ihrer Enstehung* remained a highly important tome to Mall and others interested in early human development, in the organization of embryos according to a definable scheme, and in the mechanical causation of morphometric change. His was Europe's premier embryo collector. Although his ambition to form an institute around those specimens was singular, collecting embryologic and fetal material had been considered within the realm of physicians and apothecaries since the Middle Ages.[2] Ultimately, the CIW effort embraced a considerably greater array of tasks and goals than merely collecting, as fascinating as that might have been to those who could envision a "microcosm

Figure 2.1 Franklin Paine Mall (National Museum of Health and Medicine, Armed Forces Institute of Pathology).

in a bottle" when confronted with even a single specimen.[3] The activities at the CIW also depended upon preparing the specimens for study by staging, serially sectioning, and sequentially arraying the fixed embryo sections on glass slides – a technique that matured to a significant extent in the second half of the nineteenth century in the hands of His and others. In general, the use of serial sections by zoologists and others precipitated major change in the study of development; rather than observe a single organism as it developed over time, one might evaluate morphologic change in a number of individuals representing the same organism. This change, coupled with related technological innovations, allowed emphasis on "development as a process occurring within a living organism in contact with its environment to development as a series of pictures of tissues and organs," and lead to a morphology that some grew to consider distanced from inquiries on adaptation in the individual, as is succinctly and thoroughly described by Nyhart.[4] This process, involving observing one serial section after another, is discussed as "seeing-in-time" by Keller.[5]

Observation characterized morphologic and morphometric description and organization. It dominated work using the collection. Nevertheless, the CIW's emphasis on the department as a research tool imbued the department with a liveliness that was responsive to change in technologies and key to its use.[6] Yet, for all its seemingly standardized work, this observation was built on a robust existing technology relying upon advances made by His and others: a microtome, allowing the user to cut sections of regular thinness; the early use of paraffin as an embedding medium; the use of alcohol for dehydration; a reliable process for fixation using osmium tetroxide; clearing with aromatic oil; effective mounting procedures; an embryograph (after Hartnack of Paris) for measuring and rendering images; and reconstruction techniques.[7]

When Mall published his eloquent "A Plea for an Institute of Human Embryology" in 1913, he was familiar with this arsenal and was perhaps the world's best-positioned individual to assess the advantages of an institutional setting to study early human development and the extra-institutional requirements for such an undertaking. He had come to believe that the study of embryology could effectively, and ultimately, solve the problem of congenital defects in human beings. He crystallized the role of an institute and its imperative: to establish a complete embryologic and scientific basis for human anatomy, devoid of the chaos that was currently, according to Mall, its hallmark. The collection was the imperative; paired with the work going on within it, it *was* the institute. But he also identified a basic and compelling irony about collecting. He wrote: "At present a single investigator easily makes a collection of embryos which is far beyond his power to study systematically during a lifetime, and still such a collection is too incomplete to show in a satisfactory way the continuous development of any organ." He went on to articulate the specific questions one could address only with a large collection (the curve of growth, anatomy of various stages, morphology of the brain, histiogenesis, cause of abortion, study of monsters, study of [hydatidiform] moles, and comparative and experimental embryology to elucidate human development). Mall also listed the fields that stood to gain from such studies: anatomy, physical anthropology, comparative embryology, physiology of gestation, pathology, and teratology. He skilfully cemented human embryology in the constellation of emerging data-based sciences as he compared the routine demands of image management in the study of the embryo to those in astronomy and geology.[8] He concluded that the chief function of such an institute should transcend the collecting effort itself and extend to the formation and solution of problems. Personally, his main scholarly contributions would build upon his interest in the causes of teratological abnormalities.[9]

The remainder of this chapter explores the role of the collection in this imperative over three sections: the founding of the collection and Department; the early years of the Department and the identification of major problems that would influence the use of the collection throughout its history; and an examination of special technologies applied to the collection's components in order to make them accessible to scholars and the public – three-dimensional models and public displays of work by the departmental staff. It concludes with a brief description of the collection as it passed from its Baltimore locations, the day-to-day use by Carnegie staff and the new roles it now fulfills.

Founding a collection

Even before William Welch prevailed upon Franklin Mall to leave the University of Chicago and assume the first professorship of anatomy at the Johns Hopkins Medical School in 1893, Mall was as internationally renowned for embryology as he was for his institution-building visions in medical teaching. His academic training had been influenced by the work of his teachers Carl Ludwig and Wilhelm His in Leipzig, the latter having proposed a research institute for embryology and neurology of his own.[10] Scholars debate which scientist had the greater influence on Mall; the story of those influences is well told in a number of Mall memoirs and biographical treatments, but some points bear upon Mall's interest in embryology. Otherwise, there is herein no attempt to recraft biographical material otherwise available.[11]

Mall is generally credited with having created a sound intellectual and research-based footing for the modern study of human anatomy and its pursuit as a profession both within and beyond the medical school. He influenced reformative changes in medical education throughout the 1880s and 1890s; indeed, his earlier European work exposed him to medical research initiatives he emulated throughout his career. Two related events punctuated his growing role as a leader in developmental anatomy. In 1884, Mall arrived in Leipzig and initiated his studies with Wilhelm His.[12] Under His, Mall addressed his initial embryological studies on the origin of the thymus in the human and was exposed to the demands and virtues of systematically organized, serial-sectioned embryo collections (Fig. 2.2). The next year, Mall began a working relationship with Carl Ludwig, who provided conceptual guidance regarding the nature of physiology research.

On return to the USA in 1886, Mall began his studies with William Welch at Johns Hopkins as a fellow and assistant. For three years he pursued work in bacteriology, pathology, embryology, and anatomy, initiating his first persuasive attempts to assemble an embryo collection. He left for an adjunct professorship in anatomy at Clark University, leaving Worcester after only one year, for the University of Chicago. There he exercised his growing interests in building a laboratory and a new department of anatomy, as well as constructing a program of study he felt would allow the investigation of problems of enduring significance to the biological sciences. Mall also published in this time his first major work describing a very young, seven millimeter embryo – one he had received earlier at Hopkins.[13]

Welch brought him back to Hopkins within a year. There he continued to assemble the tools of the institution – and crystallize his own vision of the dominant problems in anatomy and biology. The organized development and use of an embryo collection, as influenced by His, had significant parallels with the development and use of a research institution, as influenced by his more recent mentors.[14]

Careful observation of the morphological features of the embryo dominated work with His.[15] Using emerging microtechniques His constructed a compelling case for the position that ontogenetic events were the results of mechanical influences on differential cell growth. According to Nyhart,[16] microscopic anatomists examined well-described structures and articulated

Figure 2.2 Slide and photomicrograph, as prepared by Wilhelm His, *c.* 1860 (National Museum of Health and Medicine, Armed Forces Institute of Pathology).

the mechanical characterization of the effects of physical forces on those structures. They sought developmental laws of form. This did not compel a stance on development as a continual process or as a set of discrete stages, each with its unique anatomical descriptors, and devoid of a need for physiological explanation. Hence, the use of serial sections as prepared and assembled by His, and later Mall, altered the very practice of the study of development. Before His's development of a practical microtome, the uninterrupted study of individuals as they developed dominated practice.[17]

While at Hopkins, Mall continued to incubate the idea of a research institution devoted to embryology, as had His. Mall's involvement with academic publishing, his executive roles in academic and professional societies, his well-considered suggestions for institutions of anatomy and embryology at major academic centers,[18] his growing prominence in medical education that allowed him to cultivate professional friendships of abiding influence, and his sustained effort to assemble a collection of embryos, prepared Mall for interaction with the CIW. Mall believed a research institute must have a cornerstone, a key problem on which to focus. For Mall, it would be the development and use of a collection of embryos.[19]

The CIW provided its first grant to Mall in the spring of 1913, a modest $6,000 to organize a laboratory "of sufficient magnitude to carry on selected problems of broad scope, particularly such as are beyond the reach of a single individual." Within the year, Mall assembled at the Anatomical Laboratory of the Johns Hopkins University an investigative staff of four researchers and numerous other technicians and modelers such that by the end of 1914, Robert Woodward and the other trustees of the Institution established the activity as the Department of Embryology with Mall as its Director. It was the fulfilment of a lifetime. Mall commented, "and forever more I shall have all I need excepting grey matter."[20]

It would never be the case that all research would be based in the collection, yet the collections helped define many major problems. Work focused on organizing and expanding the collections, refining descriptions of morphologic change over time, and characterizing specific organ development. From the start, resources were made available to house Mall's collection, by then numbering over 1,000 embryos, along with their records and related material, in fireproof vaults. Mall himself described the collecting activity as the result of "unceasing efforts during the past twenty-seven years." He donated his collection to the CIW only after he became satisfied that groundwork was in place for an appropriate institute.[21] He alone had not set those stones; rather, that was a group effort that illustrates at once Mall's close relationship with Harvard's Charles Sedgewick Minot and the regard embryologists had for precious preserved and organized specimens.

For periods when Mall first established the embryology effort at Hopkins, Mall and Minot were in almost daily communication. The senior man was both mentor and sounding board for the younger; he had also studied under Ludwig, and was, like Mall, committed to improving medical education. They published in the same journals and, with George S. Huntington, had founded the *American Journal of Anatomy*. Preeminent in microscopic anatomy, Minot was likewise a superb embryologist and knew the surpassing challenges and values of a collection. His work in human embryology was recognized as the fullest expression in the field by 1892.[22] And, like Mall, he wished to establish an institute for embryology. Minot had once written to Mall regarding the goals of such institutions and the importance of differentiating them from work at the Rockefeller; obviously, debate swirled about the funding of such an enterprise. Rockefeller, felt Minot, "could tackle a specific disease entity and make progress against it. An institution [of embryology] should pursue general work in fields, rather than specific subjects." He

further noted "embryology is the pet field of research at all our laboratories and is likely to remain so."[23] Bolstered, Mall prepared a plea for an
institute of human embryology for William Welch, who was seated on the
Board of Trustees of the CIW.

In fact, Mall's ideas were the mature product of more that a decade's
thought about the roles and composition of a research institute. He had spent
a year in Carl Ludwig's Leipzig research laboratory, and had been involved in
general deliberations about the roles and needs for research institutes in the
USA through his academic positions.[24] In relaying his ideas to Minot, Mall
revealed his feeling that Carnegie should support such an institute, and that
the initial reception had been favorable. Mall had in mind a central laboratory
associated with some important medical school, or near to it, a cooperative
program, and the opportunity to extend grants for special investigations.
He felt the life of the institute should transcend the tenure of its staff and
be a "place where we can really undertake to make a great collection which
must bear upon problems in anatomy as well as in pathology. It should not
be a storehouse but it should be a live place where problems are pursued
vigorously." Minot responded immediately that Weir Mitchell would be
supportive, and reminded Mall that he and Woodward were old friends. He
pressed Mall for more information – on location and building needs. "If we
can get your new plan through, then we may bless the unborn and rejoice
together."[25]

Mall replied immediately, noting his strategy to work with Abraham
Flexner and others to bring the idea to Woodward at the CIW through
Samuel Pritchett at the Carnegie Foundation. Mall was concerned about the
Institution's interest.[26] Minot assuaged Mall's concern about what they had
come to call "their project" and distinctly approved Mall's approach through
the Foundation, to Woodward, and then directly to Carnegie, noting Welch
was important only as a backer in the project. Minot wrote, "If Carnegie
accepts their advice, the thing will go. Much should be made of the fact that
this will give Mr. Carnegie an opportunity to do something for the medical
sciences and thus show that he is not less interested than Rockefeller in this
side of philanthropy. I am very sure that I am guessing correctly. If you think
I am, you will know how to go ahead."[27]

Mall went ahead, but not without an additional consideration – joining his
mentor at Harvard, for Minot had begun to try to entice him there to chair
anatomy. In their frequent correspondence, Mall laid out his thoughts and
conditions for a serious consideration to proceed: Mall would need a significantly better salary, better staffed circumstances than he had at the anatomy
department at Hopkins, a significantly larger embryology research staff than
that which existed in Baltimore, and the inclusion of physical anthropology "in a worthy way" in any new institute.[28] He continued to explain his
efforts in his deliberations with Minot; Minot continued to remind Mall that

central to any such institution was its collection and the leadership to assure its proper use. Mall was the key to both.[29]

Woodward also wrote to Mall, indicating that he was "profoundly interested" in Mall's insistence that the research institution be separated from academic establishments (a position Woodward himself held) and assuring him that the project had a strong personal appeal,

> for embryology is one of those fields already so well worked that it may be now entered with a degree of organization and effective research quite impracticable of applications in many other fields. It has been demonstrated on a grand scale in the brief history of the Institution that it is easiest to do effective work in those fields of science already well organized and from which we may eliminate the bias of personality, creed, and nationality. Personally, I am disposed to go as far as to maintain that the best thing, even in the interests of the newer sciences, the Institution can do is to confine most of its activities to the older sciences wherein we run the least risk of wasted energies and resources in overcoming the friction of captious opposition.

He went on to ask Mall if he would take charge of such a department of research if the Institution would fund it. He concluded by asking to meet with Mall.[30]

This they did five days later, and having gained confidence in CIW's interest, Mall wrote a moving letter to Minot, indicating that it would now be unwise to leave Hopkins. He formally declined the offer the next day, in a letter marked with admiration and respect for Minot and the friendship they shared. He confessed to elation over the whole Carnegie matter, and began to describe his new concerns: "The scheme had refined much sooner than I had anticipated. In fact, it is going too rapidly to suit me, for it might abort. In order to prevent such a catastrophy, I have urged Woodward to consult with eminent embryologists (Keibel, Retzius, Minot) and this he has agreed to do . . . I must find out whether the Germans will cooperate, whether we can get the His and other collections . . ."[31] He also began to meet often with John Shaw Billings and William Welch about the actual operations of the embryology program, including definitive arrangements as to its location. Mall felt such an institute could succeed only in Berlin, Vienna, Freiburg, Boston, or Baltimore, largely because of the potential to continue to build the collections through good working relationships with gynecologists and other physicians in private practice. Clinical collaboration was critical.

As he had demonstrated the success of such a collection-building enterprise in Baltimore in connection with the Johns Hopkins Hospital and other institutions over twenty years, Mall preferred Baltimore. He went on, in a letter to Woodward, to defend his choice by describing how operations already in place there could be strengthened (but noted that space in the Anatomical Laboratory at Hopkins would be adequate for only a year before it would be necessary to construct new laboratories). Mall also prioritized his tasks: collect embryos, establish cooperation with physicians and investigators (e.g., sharing the collection, having prevailed upon the help of clinicians to assemble it), and perform research.[32]

By January of 1913, the Institution had assembled a sub-committee of its Executive Committee to determine an appropriate amount for initial funding of embryology researches. But Mall had envisioned a permanent laboratory, and although he proceeded with caution, he lost no time in continuing to emphasize the need for physical facilities and a stable staff. On the issue of a permanent department holding association with a university, Mall heard from prominent colleagues, among them Weir Mitchell, who opined that it would be impossible to found an institute for embryological research wholly within a university structure. From Simon Flexner, he heard that the institute was assured, and along with it, Mall's future as its leader – in Baltimore. By then word was out that Minot had been trying to entice Mall to Harvard. Flexner cautioned strongly against it. Mall must devote his full energies to the development of the institute rather than to any ordinary or routine work. Leading the "first embryological institute of the World," was a "'grand" opportunity, and you are the person to make a success of it."[33] Mall was indeed already deeply involved in institution building, the issues of staffing and working out the day-to-day relationships with Hopkins. His aim was a permanent department with its own building, and Mall began with plans for a busy first year of collection building, inventorying the existing materials, refining a staging mechanism for the embryos, and assembling a specific research agenda. Although Mall included mention of the tissue-based work that had been long underway at Hopkins, he emphasized collections-based, descriptive activities.[34]

Woodward urged Mall to make permanent, lasting preparations or none at all. In writing to Abraham Flexner, Mall commented: "This advice suits my convenience just now, but if I had no obligations to Hopkins I do not think that I should be inclined to follow it."[35] Mall was vexed with the space problem and the inability to gather the skilled model-makers, physical anthropologists, histotechnologists, and other technicians from the Baltimore region, but felt that as long as their space was limited, funding to work within it was more than adequate for the year 1913.[36] Still, he began his searches for staff, writing to colleagues from Europe and the USA for recommendations. He also initiated the pressure on the CIW to produce scientific papers of as superb a form as they were produced in Europe – hence requiring artists, modelers, photographers, and others.[37] That pressure was ultimately answered with the landmark publication series *Contributions to Embryology*.

His requests were informed by a watershed publication; on May 24, 1913, the *Journal of the American Medical Association* printed Mall's "A plea for an institute of human embryology" – the first "public" expression of Mall's desires and plans. He cast the document as a renewal of His's plans for such an institute, and emphasized the need for embryo collecting, particularly among the earliest stages.[38] He reviewed the paucity of existing collections and set the stage to call upon the medical community for both specimens and their records to ensure research value. Privately, however, Mall wrote to his colleague Gustav Retzius regarding his desire to have the very best investigators, supported by the very best technical aid. He also described his difficulties in acquiring the latter, particularly individuals to prepare the serial sections and make use of special stains. He wished to hire someone who had mastered physical anthropological technique to measure the embryos and tabulate those measurements. "It would be a mistake for us to make a group of measurements which would not articulate with those of anthropologists." He preferred someone trained by Retzius.[39]

In Mall's own words, collection-based work remained a core activity of the institute and one of his major contributions to science. He did not broadly disseminate his own sentiment about the effort to assemble the embryos, but came closest to it in his posthumous comments published in a memorial volume of the *Contributions to Embryology* and edited by Arthur William Meyer in 1921.

> The collection of human embryos belonging to the Carnegie Institution of Washington owes its origin to thirty years of untiring effort on the part of one of the authors (Mall). The first specimen was obtained while he was a student under Professor Welch in the Pathological Department of the Johns Hopkins University; very soon another, in excellent state of preservation, was added. After his (Mall's) transfer to Clark University in 1899 [sic; should read 1889], embryo No. 2 was studied and modeled in wax. This was the first reconstruction of a human embryo ever made in America and at that time the most elaborate one in existence. In 1890 this specimen was offered to Professor His, who refused to accept the gift, and returned it, together with several from his own collection, expressing the hope that this small number of specimens might serve as a nucleus for a much larger collection. With the subsequent foundation of the University of Chicago, the collection was transferred there, and during the following year a few additions were made. Now somewhat augmented, it was returned to Baltimore in 1893, at the opening of the Johns Hopkins Medical School, and here it grew for a number of years, at first slowly, then more rapidly, until it was finally taken over by the Carnegie Institution of Washington.[40]

Throughout the festschrift, every opportunity was taken to acknowledge appreciation for and necessity of local and more distant clinicians as collecting proceeded and as the work of the department took shape along the lines Mall described.[41] Indeed, it became a continual refrain in the pleas for donations that permeated Mall's publications and flyers from the Department, his scholarly writing, and his professional life. It is best characterized in a 1913 letter to Woodward:

You know our embryological scheme is of necessity co-operative, that is the physicians who get the specimens must co-operate with us in order that we may do our work. They are the altruistic ones, and we, the embryologists, are the selfish ones. I do not think it would be right if we remain altogether selfish, but we in turn should co-operate with other embryologists who can aid us very much in making proper use of our valuable material. You see we must be co-operative all around. In all of my experience in biological research with the possible exception of the study of eugenics, I know of no other branch where co-operation is so necessary as in the study of human embryology.[42]

On Mall's death in 1917, virtually all memoirs acknowledged that position and its hegemony as the Department progressed through the plans he had made years before.

Collecting, staging, contributions

George Streeter fulfilled Mall's dream. Even before Mall's unexpected death, George Linius Streeter (Fig. 2.3) had been identified as the next leader of the Department. As a friend and colleague, Streeter had been in constant contact with Woodward during Mall's unexpected fatal illness. Streeter dutifully reported that, two days before his death, Mall had been inquiring about laboratory bills. The day before Mall's death, Woodward appointed Streeter Acting Director of the Department; he was named Director the following month.[43]

Mall had recruited Streeter, a long-time co-investigator and then Professor of Anatomy at the University of Michigan with interests in the development of the central nervous system, as a member of the staff virtually at its inception. Streeter inherited a 6,000 square foot floor of a new, fireproof building at Hopkins, adjacent to the Anatomy Department. The floor had been specially designed for the activities of the Department of Embryology and included a library, a room for microscope work, a machine shop, and a modeling facility. There were also ten smaller rooms for investigators and their assistants, a basement room, multiple dark rooms, and a vault for the specimens and the collections.[44] As his staff expanded, he oversaw the addition of other floors in that Hunterian Laboratory Building; these were used as study rooms that housed models, reprints, books, and catalogs, as well as rooms for administration and artists' rooms. The rooms for artists in particular were of great importance; James Didusch and others were among the most superbly capable renderers of the embryonic human form and their interpretations of the CIW specimens graced hundreds of publications. In addition, Streeter inherited a growing collection that was the beneficiary of Mall's traditions of collaborating with clinicians and enticing their contributions by providing easily used standard forms and adhesive-backed stickers for mailing, and his commitment to assessing clinical information for their

Figure 2.3   George Streeter (National Museum of Health and Medicine, Armed Forces Institute of Pathology).

use in patient interactions. This collection had also attained the status of largest human embryology collection.

Like Mall, Streeter had studied with Wilhelm His and had held a position as a member of the Hopkins anatomy staff before leaving for other academic roles. His interests in three-dimensional reconstruction and the staging of embryos into "developmental horizons" marked his interests in morphology, but under his leadership, the Department made broader contributions to embryology through its publishing program and in the growth of work transcending the study of sectioned human embryos. The addition of primate colonies and the fostering of cell-culture-based investigations are examples of the broadening of the department missions.[45] In fact, part-way through his twenty-three year tenure as director of the department, Streeter assessed its major investigatory contributions: increased understanding of very early human development (and the necessary collection of human and non-human primate materials to inform it); demonstration that the developing brain and spinal cord are highly responsive to their surrounding tissues; the work by Lewis Weed and Louis Flexner on the formation of cerebrospinal fluid; identification of evidence of the autonomous nature of the movement of chromosomes during cell division; studies of the relationship of hormones obtained during uterine contraction to the knowledge of factors regulating sperm and ova transport; and cytological details of malignant cells.[46] These contributions mark the work of a laboratory led by a scientist committed to clinical collaboration. Streeter was also artful in his collaborations with the Johns Hopkins institutions, both the university and the medical school. Shortly after assuming the directorship, he provided assistance in anatomical teaching when Hopkins staff were displaced by the war effort, and throughout his tenure, he cultivated excellent working relationships resulting in useful space and resource blending.[47] He also added significantly to the investigators in the laboratories of the department. However, when faced with decisions regarding the continued employment of an investigator that Vannevar Bush, then President of the CIW, felt had become unfocused and lacking in depth, Streeter characterized his view of the scientist in this way:

> [I] pointed out his tendency to superficiality, though a liability in research, becomes an asset in the public-relations office. It seems to me that he might be happy and quite useful in such directions. If he can be used to good advantage in the Institution it would be a fortunate solution to his case. One thing I feel sure of, that a man by nature is either a good investigator or is not. If he is not, no amount of direction will make him one. Furthermore, if he is a good investigator it is probable that we should be very careful about trying to tell him what to do. Before he is taken into the Institution is the time to select him and on the basis of his self-chosen field of research and its adaptation to our own interests. After he is in, if he is not capable of feeling his own further course along profitable lines of endeavor, we should face the fact of having made a poor choice. The gardener does not teach the lily how to grow, he merely provides it with the requisite responsibility.[48]

This approach to independent thought, and its role in the work of the department, marked much of Streeter's own institution-building. He added young investigators, cultivated the attention of established ones, and pursued Mall's traditions of collecting to provide a fertile repository of material.[49] Early in the Department's history, Herbert Evans distributed circulars by the thousands to local physicians; these resulted in the offer of thousands of embryos for the collections. Mall and Evans arranged to remove any lingering concerns about the legality of the accessions by providing clarifying language from the State of Maryland's Department of Health (of which William Welch was President), which also made clear that they allowed physicians to report a stillbirth after it had been transferred to Mall's laboratory, in order that the utility of the embryo not be compromised.[50] Many of the better, usable embryos were retained and prepared immediately by Evans and others to avoid the inevitable artifacts of handling and shrinkage due to fixation.

Sectioning, staining, and mounting followed.[51] Collecting remained the most aggressive activity of the Department to support its existence as a human development enterprise. Mall and his staff never claimed that the collaborations of the chemist, the pathologist, and others were unnecessary to the enterprise of embryology, but the growing collections could form the basis of their collaborations.

Nevertheless, there would never be enough embryos. Among those that became available, their utility was sometimes compromised by a lack of complete records or due to a considerable variation of data relating to size. Inaccuracy and lack of uniformity in making measurements could invalidate, or call to question, the external morphometry that was key to staging, to basic description, and to the interpretation of anomalies that were so key to Mall's interest in the nature and causes of stillbirths and abortions.[52]

Just the same, the embryo collection rivaled almost any in size and ability to be used as a laboratory for investigation. The pacing of its development attests to the success of Mall and Evans's circulars.

> Twenty five years ago it took 10 years to collect our first hundred specimens; 5 years to collect the second hundred; 3 years for the third hundred, 2 years for the fourth hundred. And now, since it has been taken over by the Carnegie Institution of Washington, 400 specimens have been collected each year. On average, 60 physicians have contributed each 100 specimens, and the whole collection has been obtained from 509 physicians residing in 46 States and countries. I wish to emphasize again the importance of this cooperative research undertaking, in which, on account of their altruistic spirit, over 500 persons, most of them unknown to us personally, have taken part.[53]

This success is all the more remarkable in comparison to other nationally significant medical collections. At the College of Physicians of Philadelphia at the time, only twenty intact embryos could be counted among the collections. The Wistar Institute reported even fewer. At the Army Medical Museum, perhaps the largest of the human anatomical collections in the nation, the number was less than thirty.[54]

To help address the perceived paucity of materials, in 1926 Streeter secured a macaque colony, under the direction and operation of Carl Hartman, for the investigations of the physiology of mammalian reproduction and the production of very young embryos for comparative purposes.[55] With that initiative, by 1934, Streeter was able to claim that the Department had achieved

> unquestioned leadership in the assemblage and preparation of embryological specimens and associated data. This material is being examined and utilized under the most favorable conditions and the results are promptly being made available to scientists in general. Our laboratory has become a central bureau of embryological standards where an expert judgment can be obtained on developmental questions

involving normal and pathological human material . . . With such a program one cannot stand still. To have obtained a reputable position is not enough. It is our plan to continue unremittingly every effort toward enlarging and bettering our collection, and particularly to secure, through cooperation with the medical profession, additional and earlier representatives of the second and third week of development which important period is still incompletely known.[56]

The size of the collection allowed the Department to place an increasing emphasis on longitudinal studies of organ systems over time; in a sense this is a recapitulation of Keibel and Mall's work and an active acknowledgement that external morphology was only a component of the work on the horizon.

The Department employed a number of other strategies for collecting from around the globe. Mall was particularly interested in racial embryology, a term he used to categorize investigations into phenotypic differentiation based on race or demographic characteristics of the embryo. And his pleas for contributions were widely heard. Those initiatives will not be recounted here as they are well told elsewhere, but they do speak of the drive to collect.[57] Even Streeter had been moved to comment "as objects in themselves stained sections have long since lost their novelty. [It] does not mean they have ceased to be useful; on the contrary, they are securely established as a requisite tool of the investigator . . . In many phases of embryology progress can be made only through their means."[58]

When a particularly young embryo did arrive, there was cause for note, and often these objects were the celebrated focus of publications and comment in annual reports, such as the embryo collected in 1934 by Dr. Robert Tennant of Yale at a routine post-mortem examination of a 23-year-old woman who had ingested an insecticide a few hours before her death. Professor of Pathology Raymond Hussey arranged for the presentation of the embryo to the CIW collection and it was reported in a *Contribution to Embryology* under the title "The Yale embryo."[59] It is interesting to note that at the Department, the specimens often took on the names of their physician contributors, or the name of the institution with which those individuals were associated, following the practice of His.[60] This sometimes led to awkward but conventional names, such as the Edwards–Jones–Brewer ovum, another *Contribution* subject.[61]

When George Corner assumed the leadership of the Department in 1940, he described an unexcelled collection that "serves as a working museum and a kind of Bureau of Standards, which will have value as long as embryological investigators need to control their opinions by resorting to original material." He concluded that despite its status, the collection was not as complete as it could be, particularly in regard to the earliest ages.[62] It was not long before he repeated the statement and asserted in his annual report for 1946 that embryos were still needed. "Embryonic morphology has passed from the era of individual description to a phase in which group study, comparison, and statistical analysis are essential to progress. The Department therefore tends to rely upon its more constant friends among the gynecologists and obstetricians, actively cultivating the cooperation of those whose opportunities and experience make their contributions especially valuable." In the same report, he described the collection's associated materials: a collection of 795 three-dimensional wax models or graphic reconstructions, enormous file holdings, and 108 portfolios of oriented bromide prints to study when making models and from which loans were rarely made. Of the youngest embryos, loans were never extended for any reason, so precious were these materials. These conservative policies, coupled with aggressive programs to acquire young embryos, contributed to the Department's abilities to address Mall's goals, the most successful of which, in Corner's assessment, was the descriptive morphology of the human embryo.[63]

Two sets of materials are of note here, both representing materials of the greatest rarity. In 1932, Streeter had negotiated the transfer of many of Wilhelm His's embryos, together with manuscripts, instruments, and a microscope. Another group of the rarest materials, the youngest embryos of the collection, form the second set.

In 1933, Harvard's Arthur Hertig, M.D., had received a National Research Council Fellowship in support of his interests in pathological problems in obstetrics – specifically the causes of spontaneous abortions or miscarriages. He chose to spend the time in the laboratory of George Streeter at the Department of Embryology. Streeter set him onto issues in the morphology of young embryos. He learned the fundamental techniques of microscopic embryology, "the methods and values of serial sections," and how three-dimensional reconstructions could aid in the understanding of morphology by working with Chester Heuser at the laboratory. He openly attributes his interest in the earliest embryos to his time at the CIW, and in 1938 his CIW-supported search for the earliest embryos began. This is also the time Hertig had become pathologist at the Free Hospital for Women in Boston, where John Rock had begun his researches in human reproduction.[64] Hertig and Rock acquired, under Carnegie Insitution and Carnegie Foundation auspices, dozens of the earliest embryos ever seen, nearly fifty over two decades. Their work, and the celebrated occasional occurrences of the addition of a second-week embryo, peppered the annual reports of the Department of Embryology for almost twenty years. These embryos appeared in the pages of the *Contributions to Embryology* because they were noteworthy both as additions to the collection, where they received immediate, fine serial sectioned preparations, and as additions to the literature of conception and development. Even Vannevar Bush never failed to remind Streeter, and later Corner, to share with the CIW boards news about the latest early finds, extolling them as "the earliest evidence of human life" that should enjoy "a place of distinction in your unrivalled collection."[65] The Hertig and Rock embryo images were so often requested that the Institution placed image negatives with General Biological Supply House of Chicago for the preparation of "lantern slides at a reasonable cost and of course with no profit to the Carnegie Institution."[66]

In 1938, the CIW initiated the funding of a successful seven-year study on a series of 100 fertile married women on whom a therapeutic hysterectomy was performed in known calendar relation to the next expected menstrual period. Nineteen early embryos were added to the collections as a result of their work. Spurred by Streeter's support, the Department regularly funded both Rock and Hertig, and the events of the "egg hunt" are well-described elsewhere.[67] The project linked the names of Hertig and Rock with the CIW throughout the scientific world; and the Institution was well pleased with the light they shed on the relation between the embryo and the uterus, and on the timing of ovulation and of implantation. By 1952, the work of Hertig and Rock allowed Corner to describe, almost hour by hour, the behavior of the inseminated ovum.[68] He acknowledged that from the start, theirs was a joint enterprise. The preparation and study of the specimens given to Baltimore depended on the skills of Hertig and Rock. The collecting program was originally guided by Streeter and his colleagues. Further, their publications had been a joint enterprise bringing great credit to all. But when the embryo collecting slowed to a halt, as it had in 1953 and 1954, Corner encouraged them to reconcentrate any available funds on acquiring embryos of a specific early stage rather than begin to pursue biochemical investigations related to their subjects of interest.[69]

While a demographic assessment of the collections would reveal the paucity of young embryos, the practice of staging, or grouping the specimens into seriated subsets, made the lack seem glaring. Size alone was insufficient to determine the age of an embryo; other discernable features allowed researchers to describe and group standard characteristics of the embryos, and hence, to gauge the probable age of an embryo. Staging was a major project of the staff of department for most of its existence, a challenge as great with normal as it was with non-normal specimens, and staging had been a major preoccupation with embryologists since the later nineteenth century.[70] Nor was concern with staging confined to those who investigated human development. For most fields, growth curves could serve merely as guides to the assessment of morphologic change over time. And in this tendency to rely on growth curves rested a danger, as Jane Oppenheimer noted:

> The success of quantitative methods in creating the new morphology has tended to encourage attempts to adapt quantitative methods to the results of the older; and this condition, together with the fact that the journals currently encourage the publication of data in graphic and tabular form, leads to a growing tendency to make material appear quantitative which may not necessarily be so in its own right. An example is the current procedure of using morphological stage numbers from stage

series to represent ordinates or abscissae of graphs . . . A straight line in such a graph is not what it purports to be; "morphological age, as Needham ('42) calls it, is not equivalent to time, which can be quantified."[71]

Nevertheless, correlations of weight with crown–rump length, and of estimated age with the appearance of specific structures, proceeded and resulted in the tablature of distinct stages of development. Progress on describing those stages and refining those descriptions marked the pages of the *Contributions to Embryology* and the CIW *Year Books* for decades. For instance, in 1920, Streeter completed a correlation of weight with crown–rump length assessment that later informed his staging of early embryos. It represented work that had gone on at least since Mall had reported on the vexing challenges of trying to estimate an age for each embryo.[72] And before Mall, His and others had struggled with staging. It became a method that required the consideration of both external and internal states of development, and did not depend entirely on size or age.[73] Streeter became the modern champion of staging, declaring in 1922 that the Department would pursue a standardization of stages in the normal development of the human embryo. "The practice of basing such determination [of age] on the length of the specimen, which is the custom at present among anatomists," he wrote, "has proved unsatisfactory on several respects." He stated that error can derive from the posture in which the embryo is fixed, the ensuing distention or shrinkage, and the fact that the smaller specimens undergo greater relative increases with fixation, fresh specimens being more affected by the process than macerated ones. Furthermore, the greatest need for data was for embryos from the first two months of gestation.[74]

Streeter focused much of his collections-based work on staging and publishing a series of five masterful descriptions of the periods he called "horizons" in the *Contributions to Embryology*. The studies were based on measurements of over 700 embryos and occupied much of Streeter's life after his retirement from the directorship in 1940. In his introduction to the first of these, he writes:

> In searching for a suitable expression for the age groups under consideration, and one that could be used as a running title for these studies, it was decided to follow the practice of other sciences and make use of the word "horizons." The geologist, in his need for a plane of reference, speaks of a geological "horizon," meaning a stratum or associated strata characterized by particular fossils or other distinguishing features. Also archaeologists use "horizon" as a term for clarifying particular culture periods. In the present instance the term is used to emphasize the importance of thinking of the embryo as a living organism which in its time takes on many guises, always progressing from the smaller and simpler to the larger and more complex.[75]

Other publications in the *Contributions* are numerous and represent not only the work of CIW staff but that of others as well. Over time, authors of more than one fourth of the *Contributions* used the CIW collection as the basis of their work.[76] Early in his planning for an institute of embryology, Mall had envisioned regular publications from the center[77] and originally considered that they would build on his earlier work with Keibel, the *Handbook of Human Embryology* (1910 and 1912). He was prescient in anticipating the need for a central publication; publications based on the collections grew rapidly. In 1916, he reported that there had already been 141 publications based at least in part on the collection[78] and that number grew consistently. Always, the illustrations were key to the value of the volumes. Mall knew the need for the publication as a venue for papers with illustrations of the number and fineness that could not be published in other journals. As early as the first year of CIW funding, Mall pressed for the ability to publish materials of high scholarly value, even if the research had not been funded by the Institution.[79] The Institution eventually published the work, but not until Mall had received word that the Institution's Executive Committee had determined early on that due largely to the existence of the Rockefeller Institute, the CIW had determined to stay afield of medical research.[80]

Mall applied persuasive pressure. To Woodward he indicated that the Keibel collaboration was of such currency that the CIW should undertake its continuation (and contract with Keibel to remain its co-editor) as the *Manual of Human Embryology*.[81] Woodward was disinclined to confine any publications to human embryology, and the word *human* never appeared in the series title. The influential *Contributions to Embryology* were established eventually, however, and for its illustrations, Mall consistently employed the most highly skilled artists. George Corner reported that due to the superb illustrations and the thoroughness with which the standard stages of growth were illustrated, some current textbooks of human embryology acquired 25 percent or more of their illustrations from the *Contributions*.[82] However fine the illustrations, however, two-dimensional images and photographs of embryos fell short of some needs, and the Department pursued modeling and dimensional graphic representations in a way that merged historical sensitivity, artistry, and technology in new ways.

Exhibiting development

The *Contributions* and the *Year Book* served as the principal windows through which to see into the workings of the Department and learn of the nature of the scholarship accomplished there. Throughout the history of the Department, numerous requests to borrow specimens or related objects arrived from schools, universities, professional societies, and museums. Such requests were rarely granted, but very special exhibition opportunities were pursued with vigor.

In 1937, Streeter cooperated with members of the New York Museum of Science and Industry at Rockefeller Plaza by allowing their staff to *copy* models and photographs, but would not allow his staff to do the reproductions; they were to spend their time in the direction of new work, not in the duplication of existing materials. In any case, the Museum did develop an exhibition for public school children that included a glass globe representing an ovum, lantern slides showing the earliest stages of fertilization, serial section lantern slides, some plaster models, and some representative stereoscopic slides.[83] The Institution as a whole, and the department in particular, regularly participated in events for which the public was a welcome guest.

Annual publications, called *Carnegie Institution of Washington Pamphlets*, and occasional news service bulletins were produced for the lay public. The first of the pamphlets to include the new Department of Embryology described the collection and described how the staff used slides and models and drawings to understand structural anatomy better. It also introduced the concept of staging and commented on the precision of study this device brought to the work of the scientists in Baltimore.[84] Each CIW department participated in these publications, which were ghost written by the editorial staff and heavily illustrated (in contrast to the annual *Year Book*) to provide enticing views of the inner workings of CIW science. Each year's article offered some glimpse into specific research and perhaps a drawing or photographs of the facility and some staff members. Reports about the acquisition of new embryos were more than candid; they were celebrations of the unique support the CIW provided to the medical profession and to investigators in embryology and other fields. They did not obscure the nature of the work, nor did they exclude significant contributions as reported in the *Year Book*.

News service bulletins prepared by the Division of Publications appeared on occasion – major anniversaries, for instance, or for particularly newsworthy, CIW-funded activities. Three editions were printed: one for schools, one for the press, and one for staff. The two most significant bulletins to highlight the embryology effort appeared in 1933 and 1937 as part of that decade's public awareness initiative. Entitled "Human egg cells – good and bad" and "Prenatal growth of the child," they were profusely illustrated with photographs, and written in plain language devoid of scientific terminology. They were "clearing up confused thinking that has obtained in respect to the factors that control health and determine length of life."[85] The first leads with the experience of every poultryman – that not all eggs hatch a perfect chick, that some do not hatch at all, that some produce chicks of inferior quality – to introduce the concept of "good" and "bad" qualities in each egg. It goes on to include growth curves, discussions of the role of the environment and genetic factors, and a discussion of work at the CIW's Eugenics Record Office. The "Prenatal growth" bulletin is actually the transcript of an

October 21, 1936, radio broadcast that was part of a weekly series on growth and development and funded by the National Congress of Parents and Teachers. Corner's was the second broadcast of twenty-one by prominent scientists and his tone was scholarly and avuncular yet accessible, similar to that of his later reflections prepared for a popular audience entitled *Ourselves Unborn – An Embryologist's Essay on Man.*[86] He explained the growth of the embryo and fetus and addressed clinical questions on the causes of anomalies, the frequency of twinning, and the size of ova.

The CIW Headquarters also hosted annual exhibitions of departments' work during their Board of Trustees meetings throughout the 1930s. For a few days during each week, the members of the public could tour the P Street Administrative offices, where the exhibitions would be wedged into every available corner of the structure. High school students were also treated to short lectures on scientific research, in hopes that they would be inspired to consider the lessons of science in their daily lives or pursue science careers. The Department of Embryology participated fully in these exhibitions and programs, providing speakers and demonstrations about their work in model-building, staging, and photographing embryos, and about other aspects of the work in Baltimore.

Models were often the most spectacular offerings, alleviating, as they did, the challenges of perceiving morphologic change in microscopic entities. Their construction was raised to a fine art at the Department of Embryology and drew heavily and intimately on the collections themselves. Mall and Meyer reported early on that in nearly all embryology studies, it was necessary to see the structure and form of organs, ideally by observing them directly or under a microscope. But valuable specimens could not be so treated and preserved; they needed to be cut into sections and from those sections, models could be built. The process, to that date, was clearly outlined in great detail in that number in the *Contributions*, in one of a small number of methods papers to appear in those pages.[87] There, they provided an introduction to the innovations in model making that would remain a hallmark of the Department until the retirement of their chief modeler, Osborne O. Heard, in 1956 (Fig. 2.4).

Models of embryos have graced embryology research essentially since its organized beginnings,[88] but many attribute an acceleration in representation directly to microscope-related technical advances, specifically including His' approach to serial microtomy in 1866.[89] The history of model development is told well elsewhere by Hopwood;[90] what follows is an interpretation of the CIW-specific experience.

By the founding of the Department, and due to the success of his collecting efforts, Mall was able to place an increased emphasis on longitudinal studies, usually of an organ system over time. As those efforts matured, Streeter was able to report on the development of the nervous system of

Figure 2.4 Osborne O. Heard preparing specimen for mounting and modeling (National Museum of Health and Medicine, Armed Forces Institute of Pathology).

very young embryos and assemble for future study, in one place, all the serial sections along with notes, sketches, serial tracings, photographs, and three-dimensional models prepared by the master pattern maker Osborne O. Heard. For each embryo in the study, separate models were made in a uniform manner. They constituted a unit in a *series* of stages, which together came to possess a communal value. "By comparison of closely allied embryos, one can better separate the consistent and significant forms from the merely accidental." In addition, the research materials were specially arranged specifically for that purpose.[91] This kind of activity, preparing models using (and preserving) related documentation, marked one of the most enduring aspects of the Department. It expressed the collection in a sustained way that sees use today. But, however common the use of serial sections for modeling had been in other laboratories, the process had to be reinvented by the departmental staff according to its own protocols. Mall had great trouble finding experienced model makers, histologists, and some other technicians in the early years of the Department. Practical familiarity with the process was limited in the early 1910s and Osborne O. Heard, a young engineer's pattern maker who was recruited by Mall from a local art institute in 1913, began modeling afresh.

Figure 2.5 Collections storage and study in vaulted room at the Baltimore facility, c. 1950.

True to Mall's usual form, he would give Heard a problem – model this system over time – and Heard was free to proceed, creating innovation after innovation to succeed. He introduced a broadening array of materials for the final form of the models – paper, wax, plaster, acetate sheets – learning metallurgy in the process, cutting glass for slides, inventing storage techniques for their safe keeping, introducing a variety of photographic apparata and processes into the group, and calling each model "a study." He viewed his contributions among the scientific assets of the Department. Models were made to illustrate specific points as well as to create a vocabulary of form. Occasionally, he worked directly with the scientific staff of the Department and they were generous in their praise of Heard and for the insight his models offered in the process of understanding change in three and four dimensions. Often, his completed plaster model series took years to come to fruition.[92] Heard's daughter, Martha Heard Ray, recalled that Heard and Alice Caspari, a long-time departmental secretary, often worked in great secrecy and placed the models and research notes in the vault each night, where they joined the protected embryos[93] (Fig. 2.5).

Streeter viewed this work as noteworthy. In a highly unusual use of illustration in the *Year Book*, he included a drawing of a unique instrument used for measuring embryos in fluid, deeming it of sufficient interest and complexity to warrant special notice.[94] Over time, other innovations received special mention in the *Year Book*, among them a vertical stereo camera for use in rapid but precisce embryo photography and designed by the Department's C. H. Heuser and Osborne O. Heard.[95] Vannevar Bush himself got into the act, designing devices for stereoscopic viewing of opaque specimens as though they were transparent. Inveterate inventor that he was, Bush called them superposed specimens and prevailed upon the shops of the Carnegie's Department of Terrestrial Magnetism in Washington to make the complex device.[96] He followed a rich tradition of technical developments to facilitate model making and graphic reconstruction initiated largely by Heard. As early as 1915, W. Lewis had invented a "line guide" registration process for stacking images and wax in the appropriate orientation for modeling. Heard refined and built upon that process by introducing photographic apparatus. The models, and their enabling technologies, remained an intellectual mainstay of the Department for decades[97] (Fig. 2.6).

Moving the collection

In 1912, as Mall was progressing in his efforts to realize his vision of an institute of embryology, he had received approbation from colleagues in tune with his labors. From the University of Chicago Department of Anatomy, his colleague and fellow developmental anatomist George Bartlemez wrote, "I hope it will not be long before it will be necessary for every thorough piece

Figure 2.6 File arrays of CIW Human Embryology Collection, showing investigators and custom-made storage boxes, c. 1950.

of work on human development to be checked up and confirmed by a study of your series. Then the startling, half-baked conclusions we often see will be eliminated and the work made much more convincing."[98]

Never before Mall's attempt to create an institutional setting, supported as it was by the CIW specifically for the study of human embryology, did an organized, nationally based and internationally subscribed collecting effort succeed. And never has it been surpassed. But by the last years of Corner's tenure as Director, the supremacy of the position of the collection had begun to erode. Corner noted in his last *Year Book* entry that under the next Director, biochemistry would yield great progress in the understanding of development particularly with respect to the building of tissues. Yet, that biochemistry would inform a constant review of morphology, and that morphology would forge the next questions about form and structure. "The Department of Embryology will of necessity have to consider, as far as its means and staff can reach, the full gamut of life structure from the electron to the whole organism."[99]

James Ebert, the young zoologist chosen to lead the Department next, inaugurated his work with the CIW by acknowledging a resurgence of interest in the mechanisms of development.[100] However, the collection itself did not remain associated with, or even collocated with, the departmental Johns Hopkins facilities in Baltimore. His concerns about the relationship between the Institution and Hopkins are never more apparent than in his deliberations about the need for a new facility for the former. After consideration of the scientific program, he had concluded that while the staff did not need to grow beyond its numbers, that staff needed new and enlarged space within which to do its work. He favored a Baltimore site, as it facilitated informal ties with the university. He wanted a site independent of both the medical and Homewood campuses and saw only one serious disadvantage of physical independence: loss of easy library access. However, if it were necessary to locate near a Hopkins facility, Ebert preferred that of the medical school. Changes in the medical school's staff made a union more reasonable than it had been, as did a growing biophysics group. Ebert had "misgivings about the power politics" of the biology groups and their laboratories at the other campus, and he wished to steer clear of them in so far as he could, retaining a "friendly, 'distant relative'" relationship with the medical school[101] (Fig. 2.7).

By the 1960s, James Ebert had had five years at the helm of the Department and had established his directorship in accordance with the CIW's desires that he bring to their Department the newest exploration technologies, molecular studies, and a science that was not dependent upon the collections based on serial sections that had been a key to the progress of the six earlier decades. The departmental labs and personnel had moved to a new building on the Hopkins campus, one designed to support the emerging agendas of the

Figure 2.7   CIW Department of Embryology staff, 1956.

Institution.[102] Biochemistry and microbiology, along with other laboratory-based pursuits such as immunology and radiology, were now informing the mechanisms of human and comparative development, cell differentiation and diversification. The basic tenets of work at the laboratory had not changed; reproductive sciences continued to attract original and independent thinkers who pursued their investigations under the CIW imprimatur.

With the retirements of both technicians and the researchers who had helped to cultivate the collections and who were intimately familiar with the portrait of development each embryo represented, fewer and fewer scientists routinely relied upon the collections to perform their investigations. In 1957, the Institution decided to remove the collection to an ancillary position and ultimately rehoused it at the California Primate Research Center of the University of California at Davis, California, where it reopened in 1975. The actual physical mothballing of the collection in 1971 to prepare for a move to Davis had officially curtailed most of the CIW's human embryology program.

At Davis, Ronan O'Rahilly and Ernest D. Gardner maintained associate academic appointments at the university while they carefully curated the collection and allowed some researchers access to the holdings. Still operating under the title of the Carnegie Laboratories of Embryology, they continued their own work on the development of the nervous system in the earliest human stages. O'Rahilly attempted to organize a major catalog of all major human embryology collections world wide, and beginning in 1980 relied upon the efforts of embryologist Alexander Barry to compile such a tool in order to demonstrate the primacy of the Carnegie Collection and secure support for an International Secretariat for Human Embryological Material.

O'Rahilly failed to secure the secretariat. However, several of the collections listed in the document he and Barry had hoped would be persuasive, regarding the need for such an organization in light of the dispersion of collections, are now collocated with the Carnegie Collection: the Hooker–Humphrey collection at the University of Alabama; other Davis collections; and at the time of this writing two others are pending.[103]

Before his retirement, O'Rahilly assembled a thorough literature collection as well and, along with Fabiola Muller, published a masterful description of each of the twenty-three stages that represent the first eight weeks of human gestation.[104] Theirs was the culmination of the work of their CIW predecessors – careful descriptions, superb illustrations and photographs, and a sensibility clearly informed by the three-dimensional models that artists and artisans of the CIW had created over the years in Baltimore.

With the conclusion of his most active period shepherding the collection, Ronan O'Rahilly remained on the CIW staff but returned to Europe. By early 1990, the Institution faced a decision about the fate of the collection itself. Leaders of the Institution and the Embryology Department wished to keep the collection intact and available to qualified scholars, to keep it in the continental USA, and to keep it from becoming the proprietary asset of a single university. Accordingly, after a number of organizations had expressed their interests in providing an institutional home for the material, the Institution selected the National Museum of Health and Medicine of the Armed Forces Institute of Pathology in Washington, DC. It was committed to broad-based, safe access to the slides and public availability of a collection while retaining its historical integrity. By the time it was under serious consideration as a repository for the Carnegie Collection in 1990, the museum had developed a reputation for fostering responsible research on non-renewable landmark resources.

The Carnegie Collection arrived in Washington in 1991 with all 10,299 associated records (one for each individual that had ever been in the collection), whole embryos, models, instruments, publications, and thousands of reprints carefully collected and organized by O'Rahilly and Gardner. Elizabeth Ramsey once again provided guidance on its use (Fig. 2.8). Since then, it has hosted over three hundred researchers from the world over who use it for a widening array of purposes, some of which would have been familiar to Mall. Examples are the earliest appearance of particular morphological features, the refinement of stage definitions, the simultaneous expression of internal and external features, teratological features, and the like.[105] Some wish to apply emerging imaging technologies to long-preserved intact embryos. Others are as interested in the challenges of organizing, preserving, and sharing the collection contents in an automated format (which, the reader will recall, Mall anticipated in his 1913 "Plea", recognizing the needs of information management).[106]

Figure 2.8   Elizabeth Ramsey at her home in Washington, DC, *c.* 1990 (National Museum of Health and Medicine, Armed Forces Institute of Pathology).

Most recently, the collection has experienced a renewed public awareness and is available to anyone who wishes to see it in the virtual setting of a website or electronic file transfers. Beginning in 1992 and lasting until 1997, the National Institute of Child Health and Human Development led an effort joined by the National Center for Research Resources and the Office for Research on Women's Health to make the collections electronically accessible. Inventories were created, and, not unlike the earliest efforts to make four-dimensional change appreciable to scholars and clinicians, computer models were made, each one highlighting up to six organ systems. Now they appear as slightly crude examples of the computer model maker's art, the technique replicating on the computer screen the wax-based process of Born and later Osborne O. Heard. Electronic development and distribution of these models minimizes damage to the actual models, but when nothing substitutes for seeing the model or for peering through a microscope to see a specific feature, the museum hosts visitors to the collection (Fig. 2.9).

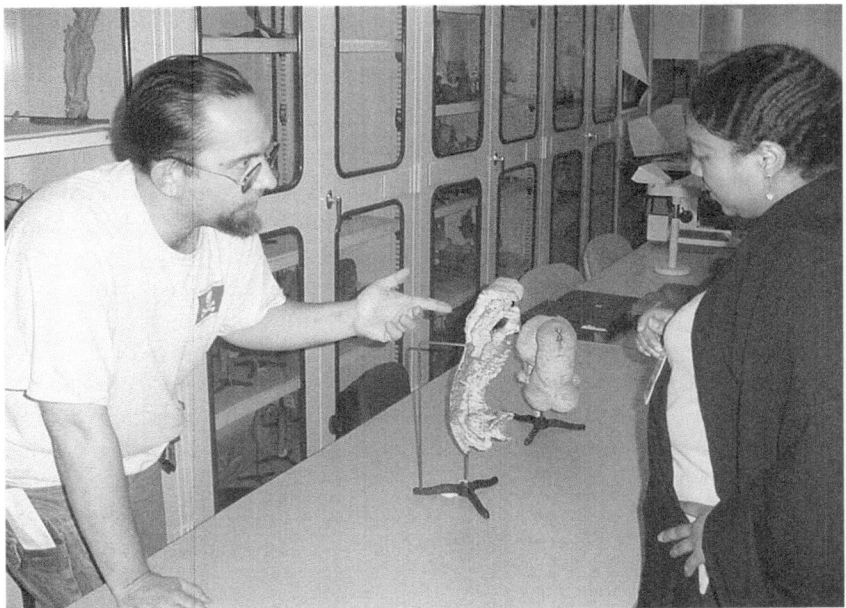

Figure 2.9   William F. Discher and Surinder Sandhu of the National Museum of Health and Medicine of the Armed Forces Institute of Pathology examine Carnegie models, 2003 (National Museum of Health and Medicine, Armed Forces Institute of Pathology).

A subset of the materials resides on public display in the museum now in an exhibition that evokes the public exhibitions of the CIW Administration Building of the 1920s, 1930s, and 1940s. The displays are complete with models, explanations of the modeling process, and photographs of the embryos. They are paired with a new picture-based tome on human developmental anatomy.[107]

The most ambitious computer-based project to date is a National Library of Medicine Next Generation Internet Grant to a collaboration of partners assembled to use the collection in educational, clinical, and research settings. Using a sophisticated real-time distributed network of networks, participants from the George Mason University, the National Museum of Health and Medicine, the Johns Hopkins Medical Institutions, Eolas Technologies, Inc., the University of Illinois Medical School in Chicago, the Lawrence Livermore National Laboratory, the San Diego Super Computing Center, and the Oregon Health Sciences University are developing massively interactive data sets for use in any environment. Ever loyal to Mall's views, in 1920 Streeter succinctly described the purpose of the collection in the form of stained sections: to constitute "a great storehouse of data from which the story of the development of the body is gradually being revealed."[108] Today, they continue to fulfill many of Mall's original visions for an institutional setting for the study of human development.

Acknowledgements

The author wishes to thank, with gratitude, Maxine Singer, Jane Maienschein, and Garland Allen, and wishes to acknowledge the help of Elizabeth C. Lockett, William F. Discher, and Kumudini Mayur of the staff of the Human Developmental Anatomy Center of the National Museum of Health and Medicine of the Armed Forces Institute of Pathology; the staff of the Carnegie Institution of Washington, particularly Margaret Hazen and John Strom, and Tina McDowell; and Gerard Shorb and his colleagues at the Alan Mason Chesney Medical Archives of the Johns Hopkins Medical Institutions. In addition, she is indebted to Donald Brown, Adele Clarke, Michael Doyle, William A. Gardner, Raymond Gasser, Marie Glitz, Martha Heard Ray, Robert J. T. Joy, Donald West King, Lynn Morgan, Glenn N. Wagner, and the late Hans A. Klagsbrunn and Elizabeth Maplesden Ramsey.

Notes

1. Ramsey, Elizabeth M. (1938), Untitled poem, Carnegie Files, Human Developmental Anatomy Center, National Museum of Health and Medicine, Armed Forces Institute of Pathology, Washington, DC (hereinafter NMHM).

2. Loraine Daston and Katharine Park, *Wonders and the Order of Nature 1150–1750* (New York: Zone Books, 1998), pp. 149, 152; and Oleg Neverov, "'His Majesty's Cabinet' and Peter I's Kunstkammer," in O. Impey and A. MacGregor (eds.), *The Origins of Museums: The Cabinet of Curiosities in Sixteenth and Seventeenth Century Europe* (London: House of Stratus, 2001), pp. 71–9.

3. Frank Gonzalez-Crussi, *Suspended Animation: Six Essays on the Preservation of Bodily Parts* (New York: Harcourt Brace and Company, 1995), pp. 1–22.

4. Lynn K. Nyhart, *Biology Takes Form: Animal Morphology and the German Universities, 1800–1900* (University of Chicago Press, 1995), pp. 202–3.

5. Evelyn Fox Keller, *Making Sense of Life: Explaining Biological Development with Models, Metaphors, and Machines* (Cambridge, MA: Harvard University Press, 2002), p. 216. Keller cites Hannah Landecker, "Technologies of living substance: tissue culture in twentieth century biomedicine." Doctoral Thesis, Massachusetts Institute of Technology, 1999.

6. On collections as tools, see Adele E. Clarke, "Research materials and reproductive science in the United States, 1910–1940," in Susan Leigh Starr (ed.), *Ecologies of Knowledge: Work and Politics in Science and Technology* (Albany, NY: State University of New York Press, 1995), pp. 183–225. On Mall and the building of the CIW human embryology collection, see pp. 195–201.

7. C. W. Bodemer, "The biology of the blastocyst in historical perspective," in R. J. Blandau (ed.), *The Biology of the Blastocyst* (University of Chicago Press, 1971), p. 18.

8. Franklin P. Mall, "A plea for an institute of human embryology," *Journal of the American Medical Association* 60 (1913), p. 1601.

9. Franklin P. Mall, "A study of the causes underlying the origin of human monsters," *Journal of Morphology* 19 (1908) p. 1–367.

10. Franklin P. Mall, "Wilhelm His, his relation to institutions of learning," *American Journal of Anatomy* 4 (1905), p. 153; Florence R. Sabin, *Franklin Paine Mall, The Story of a Mind* (Baltimore, MD: Johns Hopkins Press, 1934), p. 287.

11. On Mall, see Florence R. Sabin, *Franklin Paine Mall, The Story of a Mind* (Baltimore, MD: Johns Hopkins Press, 1934), esp. pp. 281–310; Florence R. Sabin, "Franklin Paine Mall: A review of his scientific achievement," *Science*, N.S., 47 (1918), pp. 254–61; George W. Corner, "Franklin Paine Mall," *Dictionary of Scientific Biography* 9 (1974), pp. 55–8; G. C. Huber, "Franklin Paine Mall, 1862–1917," *Anatomical Record* 14 (1918), pp. 3–17; Howard A. Kelly and Walter Burrage, "Franklin Paine Mall," *Dictionary of American Medical Biography* (Boston, MA: Milford House, 1928), pp. 803–4; A. McGehee Harvey, "A new school of anatomy: the story of Franklin P. Mall, Florence R. Sabin and John B. MacCallum," *Johns Hopkins Medical Journal* 136 (1975), pp. 83–94; Florence R. Sabin, "Franklin Paine Mall, 1862–1917," *Biographical Memoirs* 16 (Washington, National Academy of Sciences, 1936), pp. 65–122; and William H. Welch, *et al.*, "Memorial service in honor of Franklin P. Mall, Professor of Anatomy, Johns Hopkins University, 1893–1917," *Johns Hopkins Medical Bulletin* 29 (1918), pp. 109–23. See also Nancy Medley, "The uses of women: Franklin Mall and human embryology, 1900–1917." Paper presented to the Department of the History of Science, Medicine and Technology, Johns Hopkins University, 11 April 2002, pp. 1–33.

12. Sabin, *Franklin Paine Mall*, pp. 36–39.

13. Franklin P. Mall, "A human embryo twenty-six days old," *Journal of Morphology* 5 (1891), pp. 459–80.

14. Nick Hopwood, "Producing development: the anatomy of human embryos and the norms of Wilhelm His," *Bulletin of the History of Medicine* 74 (2000), pp. 29–74. See also Adele Clarke, "Embryology and the rise of American reproductive sciences, circa 1910–1940," in Keith R. Benson, Jane Maienschein and Ronald Rainger (eds.), *The Expansion of American Biology* (New Brunswick, NJ: Rutgers University Press, 1991), pp. 107–32; and, in general, Jane Maienschein *Transforming Traditions in American Biology, 1880–1915* (Baltimore, MD: Johns Hopkins University Press, 1991).

15. Jane Maienschein, "The origins of Entwicklungsmechanik," in Scott F. Gilbert (ed.), *A Conceptual History of Modern Embryology* (New York: Plenum Press, 1991), pp. 43–61; Ronan R. O'Rahilly, "One hundred years of human embryology," in Harold Kalter (ed.), *Issues and Reviews in Teratology* 4 (1988), pp. 81–128; and Mall, "Wilhelm His," pp. 139–161.

16. Nyhart, *Biology Takes Form*, p. 87.

17. Brian Bracegirdle, *A History of Microtechnique*, 2nd edn. (Lincolnwood, IL: Science Heritage, Ltd., 1987), p. 131.

18. Sabin, *Franklin Paine Mall*, pp. 295–6.

19. Sabin, *Franklin Paine Mall*, p. 284.

20. F. P. Mall to Simon Gage, 20 December 1914, Alan Mason Chesney Medical Archives of the Johns Hopkins Medical Institutions (hereinafter Chesney Archives), Mall Papers, Correspondence.

21. Franklin P. Mall to Robert S. Woodward, 21 March 1914, CIW Archives, Embryology/Director Correspondence file. See also CIW Pamphlet collection, 1915, p. 17. The effort to assemble that collection for the early years of the department is well told in Medley, "The uses of women"; for descriptions of other collecting efforts and "networks" of collections, see Lynn M. Morgan, as revised, 2001, "Embryo tales," in Sarah Franklin and Margaret Lock (eds.), *Remaking Life & Death: Towards an Anthropology of the Biosciences* (Santa Fe, NM: SAR Press, forthcoming); and Lynn M. Morgan, "'Properly disposed of': a history of embryo disposal and the changing claims on fetal remains," *Medical Anthropology*, 21 (2002), pp. 247–74.

22. Frederic T. Lewis, "Charles Sedgwick Minot," in Howard A. Kelly and Walter Burrage, *Dictionary of American Medical Biography* (Boston: Milford House, 1928), p. 850; Sabin, *Franklin Paine Mall*, pp. 264–5.

23. C. S. Minot to F. P. Mall, 11 March 1909, Chesney Archives, Mall Papers, Correspondence.

24. Sabin, *Franklin Paine Mall*, p. 294.

25. F. P. Mall to C. S. Minot, 31 July 1911, and C. S. Minot to F. P. Mall, 3 August 1911, Chesney Archives, Mall Papers, Correspondence.

26. F. P. Mall to C. S. Minot, 5 August 1911, Chesney Archives, Mall Papers, Correspondence.

27. C. S. Minot to F. P. Mall, 12 December 1911, Chesney Archives, Mall Papers, Correspondence.

28. F. P. Mall to G. S. Minot, 16 February 1912, Chesney Archives, Mall Papers, Correspondence.

29. C. S. Minot to F. P. Mall, 28 April 1913, CIW Archives, Embryology/Director Correspondence file.

30. R. Woodward to F. P. Mall, 20 May 1912, Chesney Archives, Mall Papers, Correspondence.

31. F. P. Mall to G. S. Minot, 25 and 26 May 1912, Chesney, Mall Papers, Correspondence. The suggestion of Franz Keibel brought Woodward to another Mall collaborator. With Keibel, he published in 1910 and 1912 an atlas of human embryology some believe even today has not been superseded for quality and clarity.

32. F. P. Mall to R. Woodward, 9 December 1912, CIW Archives, Embryology/Director Correspondence file.

33. S. W. Mitchell to F. P. Mall, 29 January 1913, CIW Archives, Embryology/Director Correspondence file; S. Flexner to F. P. Mall, ND, CIW Archives, Embryology/Director Correspondence file.

34. F. P. Mall to W. H. Welch, 18 February 1913, Chesney Archives, Mall Papers, Correspondence.

35. F. P. Mall to A. Flexner, 30 April 1913, Chesney Archives, Mall Papers, Correspondence.

36. F. P. Mall to C. S. Minot, 19 March 1913; F. P. Mall to Winslow Upton, 14 May 1913, Chesney Archives, Mall Papers, Correspondence.

37. F. P. Mall to R. Woodward, 24 May 1913, CIW, Embryology/Director Correspondence file.

38. Mall, "A plea for an institute," p. 1600.

39. F. P. Mall to G. Retzius, 11 April 1913, Chesney Archives, Mall Papers, Correspondence.

40. Franklin Paine Mall and Arthur William Meyer, "Studies on abortuses: a survey of pathologic ova in the Carnegie embryological collection," *Contributions to Embryology* 7 (1921), p. 13.

41. For day-to-day curatorial practices of the Department, see particularly Medley, "The uses of women"; Lynn M. Morgan, "Materializing the fetal body, or, what are those corpses doing in biology's basement?," in Lynn M. Morgan and Meredith W. Michaels (eds.), *Fetal Subjects, Feminist Positions* (Philadelphia, PA University of Pennsylvania Press, 1999); Morgan, "Embryo tales"; and Morgan, "Properly disposed of," pp. 247–50.

42. F. P. Mall to R. Woodward, 13 September 1913, CIW Archives, Embryology/Director Correspondence file.

43. George L. Streeter to R. Woodward, 15 and 17 November 1917 and 18 December 1917, CIW Archives, Embryology/Director Correspondence file.

44. Franklin P. Mall, Carnegie Institution of Washington *Year Book* 17 (1917–18), p. 99.

45. "George Linius Streeter," *Journal of Anatomy* 83 (1949), pp. 51–2.

46. George L. Streeter to John Merriam, 8 November 1933, CIW Archives, Embryology/Director Correspondence file.

47. George L. Streeter to Robert Woodward, 23 September 1918, CIW Archives, Embryology/Director Correspondence file.

48. George L. Streeter to Vannevar Bush, 28 February 1939, CIW Archives, Embryology/Director Correspondence file.

49. Medley, "The uses of women," pp. 10–32.

50. Frederic V. Beitler, Chief, Bureau of Vital Statistics, State of Maryland Department of Public Health, to Franklin P. Mall, 12 December 1916, CIW Archives, Embryology/Director Correspondence file.

51. Mall and Meyer, "Studies on abortuses," p. 28.

52. Hopwood, "Producing development," throughout.

53. Franklin P. Mall, Carnegie Institution of Washington *Year Book* 16 (1916–17), pp. 109–10.

54. The author thanks Gretchen Worden, Director of the Mutter Museum of the College of Physicians of Philadelphia and Nina Long, Archivist of the Wistar Archives of the Wistar Institute for this information. The Wistar data derives from its *Director's Reports*, v. 1912–23, Philadelphia, Pennsylvania.

55. See Clarke, "Research materials," pp. 183–225, for more on colony development and comparative investigations.

56. George Streeter Manuscript, 1934, CIW Archives, Embryology/Director Correspondence file, pp. 1–2.

57. Clarke, "Research materials"; Medley, "The uses of women"; Morgan, "Materializing the fetal body," pp. 43–60; Morgan, "Embryo tales"; and Morgan, "Properly disposed of."

58. George L. Streeter, Carnegie Institution of Washington *Year Book* 30 (1930–1), p. 3.

59. Elizabeth M. Ramsey, "The Yale embryo," *Contributions to Embryology*, (Carnegie Institution of Washington, 1938), pp. 67–84.

60. Hopwood, "Producing development," p. 39.

61. John I. Brewer, "A human ovum in the bilaminar blastodisc stage (The Edwards–Jones–Brewer ovum)," *Contributions to Embryology* 27. (Carnegie Institution of Washington, 1938), p. 85.

62. George W. Corner, Carnegie Institution of Washington *Year Book* 40 (1940–1), p. 188.

63. George W. Corner, Carnegie Institution of Washington *Year Book* 46 (1946–7), p. 129.

64. Arthur W. Hertig, *Human Trophoblast* (Springfield, IL: Charles C. Thomas, 1968), pp. 8–11.

65. Vannevar Bush to George Streeter, 12 January 1939, CIW Archives, Embryology/Director Correspondence file; Vannevar Bush to George Corner, 22 April 1942, CIW Archives, Embryology/Director Correspondence file.

66. George W. Corner, Carnegie Institution of Washington *Year Book* 43 (1943–4), p. 120.

67. Clarke, "Research materials," pp. 198–201; Loretta McLaughlin, *The Pill, John Rock, and the Church* (Boston, MA: Little, Brown, 1982), pp. 59–66; Morgan, "Materializing the fetal body," pp. 51–2.

68. George W. Corner, Carnegie Institution of Washington *Year Book* 52 (1952–3), p. 161–2.

69. George W. Corner to Arthur Hertig, 13 January 1954, CIW Archives, Embryology/Director Correspondence file.

70. Hopwood, "Producing development," particularly for the influence of Keibel's *Normentafeln*. See also F. Keibel and C. Elze, *Normentafeln zur Entwicklungsgeschichte des Menschen* (Jena: Fisher, 1908).

71. Jane M. Oppenheimer, *Essays in the History of Embryology and Biology* (Cambridge, MA: MIT Press, 1967), pp. 185–7.

72. George L. Streeter, Carnegie Institution of Washington *Year Book* 20 (1920–1), p. 83.

73. O'Rahilly, "One hundred years," p. 101.

74. George L. Streeter, Carnegie Institution of Washington *Year Book* 22 (1922–3), p. 76.

75. George L. Streeter, "Developmental horizons in human embryos. Description of age group XI, 13 to 20 somites, and age group XII, 21 to 29 somites," *Contributions to Embryology* 30 (1942), p. 213. See also years 1945, 1948, 1949, and 1951 for subsequent studies in the series.

76. For this assessment, staff of the Human Developmental Anatomy Center of the National Museum of Health and Medicine evaluated all *Contribution* entries from 1915 to 1966. I am grateful for their help.

77. Mall, "A plea for an institute," p. 1600.

78. Franklin P. Mall, Carnegie Institution of Washington *Year Book* 16 (1916–17), p. 110.

79. Franklin P. Mall to R. Woodward, 29 March 1914, CIW Archives, Embryology/Director Correspondence file.

80. [Author unnamed] to Franklin P. Mall, 4 April 1913, CIW Archives, Embryology/Director Correspondence file.

81. Franklin P. Mall to R. Woodward, 10 September 1913, CIW Archives, Embryology/Director Correspondence file.

82. George W. Corner, Carnegie Institution of Washington *Year Book* 55 (1955–6), p. 190.

83. George Streeter to K. G. Frank, 20 March 1937, Chesney Archives, Corner Correspondence, New York Museum of Science and Industry file.

84. Franklin Paine Mall, Carnegie Institution of Washington Pamphlet, CIW Archives, pp. 17–18.

85. Carnegie Institution of Washington *News Service Bulletin* (1933), np; and George L. Streeter, "Prenatal growth of the child," *News Service Bulletin* (Staff Edition) 4 (1937) Number 14, pp. 126–32.

86. George W. Corner, *Ourselves Unborn – An Embryologist's Essay on Man*, Dwight Harrington Terry Lectures (New Haven, CT: Yale University Press, 1944).

87. Mall and Meyer, "Studies on abortuses," pp. 30–3.

88. Barbara Maria Stafford, *Body Criticism: Imagining the Unseen in Enlightenment Art and Medicine* (Cambridge, MA: MIT Press, 1991), p. 21. See Karen Newman, *Fetal Positions: Individualism, Science, Visuality* (Stanford, CA: Stanford University Press, 1996), pp. 70–81 for a carefully selected array of embryo and fetal images and models interpreted from an art historical and social perspective.

89. Bracegirdle, *History of Microtechnique*, pp. 129–30; Nyhart, *Biology Takes Form*, pp. 201–2; O'Rahilly, "One hundred years," pp. 85–6; Keller, *Making Sense of Life*, pp. 215–19; Arthur William Meyer, *The Rise of Embryology* (Stanford University Press, 1939), pp. 238–77; Steve Gilbert, Correspondence with author, December 2002. MS by Steve Gilbert, *Images of Embryos, An illustrated Sourcebook in the History of Descriptive Embryology*, forthcoming, pp. 238, 241.

90. Nick Hopwood, *Embryos in Wax: Models from the Ziegler Studio* (Cambridge: Whipple Museum of the History of Science, Cambridge University and Bern: University of Bern, 2002).

91. George L. Streeter, Carnegie Institution of Washington *Year Book* 26 (1926–7), pp. 4–5.

92. Osborne O. Heard, *Recollections–Reflections, Hayden Lake, Idaho* (1979), MS, Human Developmental Anatomy Center, National Museum of Health and Medicine, Armed Forces Institute of Pathology, Washington, DC, pp. 2–3. The plaster models have been saved and are among the collections of the Human Developmental Anatomy Center, National Museum of Health and Medicine, Armed Forces Institute of Pathology, Washington, DC.

93. Martha Heard Ray, Interview with William F. Discher, December 1998. Human Developmental Anatomy Center, National Museum of Health and Medicine, Armed Forces Institute of Pathology, Washington, DC.

94. George L. Streeter, Carnegie Institution of Washington *Year Book* 20 (1920–1), p. 84.

95. George W. Corner, Carnegie Institution of Washington *Year Book* 43 (1943–4), p. 169.

96. Vannevar Bush to George W. Corner, 14 April 1949, CIW Archives, Embryology/Director Correspondence file.

97. Lewis, "Charles Sedgwick Minot," pp. 849–51; Mall and Meyer, "Studies on abortuses," p. 32; Osborne O. Heard, "A photographic method of orienting serial sections for reconstruction," *Anatomical Record* 49 (1931), pp. 59–70.

98. George W. Bartlemez to Franklin P. Mall, 10 August 1912, Chesney Archives, Mall Papers, Correspondence.

99. George W. Corner, Carnegie Institution of Washington *Year Book* 55 (1955–6), p. 192.

100. James D. Ebert, Carnegie Institution of Washington *Year Book* 57 (1957–8), p. 269.

101. James D. Ebert to Caryl P. Haskins, 12 February 1958, CIW Archives, Embryology/Director Correspondence file.
102. James Trefil and Margaret Hindle Hazen, *Good Seeing, A Century of Science at the Carnegie Institution of Washington 1902–2002* (Washington, DC: The Joseph Henry Press, 2002), p. 176.
103. O'Rahilly, "One hundred years," p. 112.
104. Ronan O'Rahilly and Fabiola Muller, *Developmental Stages in Human Embryos Including a Revision of Streeter's "Horizons" and a Survey of the Carnegie Collection*, Publication 637 (Carnegie Institution of Washington, 1987).
105. Some of this work had been anticipated by others. See, for example, Ronan O'Rahilly, F. Muller, G. M. Hutchins, and G. W. Moore, "Computer ranking of the sequence of appearance of 100 features of the brain and related structures in staged human embryos during the first 5 weeks of development," *American Journal of Anatomy* 180 (1984), pp. 69–86.
106. Mall, "Plea for an institute," p. 1601.
107. Alexander Tsiaras, *From Conception to Birth, A Life Unfolds* (New York: Doubleday, 2002). Tsiaras dedicated the volume to the Carnegie Institution of Washington for its investment in the embryology collection and in celebration of its centennial.
108. George L. Streeter, Carnegie Institution of Washington *Year Book* 20 (1920–1), p. 3.

Bibliography

Bodemer, C. W., "The biology of the blastocyst in historical perspective," in R. J. Blandau (ed.), *The Biology of the Blastocyst* (Chicago, IL: University of Chicago Press, 1971), pp. 1–25.

Bracegirdle, Brian, *A History of Microtechnique*, 2nd edn. (Lincolnwood, IL: Science Heritage, Ltd., 1987).

Brewer, John I., "A human ovum in the bilaminar blastodisc stage (The Edwards–Jones–Brewer Ovum)," *Contributions to Embryology* 27 (Washington, DC: Carnegie Institution of Washington, 1938), pp. 85–93.

Carnegie Institution of Washington, Division of Publications, "Human egg-cells – good and bad," *News Service Bulletin* (Staff Edition) 3, Number 7 (1933), pp. 44–9.

Clarke, Adele E., "Embryology and the rise of American reproductive sciences, circa 1910–1940," in Keith R. Benson, Jane Maienschein, and Ronald Rainger (eds.), *The Expansion of American Biology* (New Brunswick, NJ: Rutgers University Press, 1991), pp. 107–32.

"Research materials and reproductive science in the United States, 1910–1940," in Susan Leigh Star (ed.), *Ecologies of Knowledge: Work and Politics in Science and Technology* (Albany, NY: State University of New York Press, 1995), pp. 183–225.

Corner, George W., Carnegie Institution of Washington *Year Book* 40 (1940–1).

Carnegie Institution of Washington *Year Book* 43 (1943–4).

Ourselves Unborn – An Embryologist's Essay on Man. Dwight Harrington Terry Lectures (New Haven, CT: Yale University Press, 1944).

Carnegie Institution of Washington *Year Book* 46 (1946–7).

Carnegie Institution of Washington *Year Book* 52 (1952–3).

Carnegie Institution of Washington *Year Book* 55 (1955–6).

Daston, Loraine and Katharine Park, *Wonders and the Order of Nature 1150–1750* (New York: Zone Books, 1998).

Ebert, James D., Carnegie Institution of Washington *Year Book* 57 (1957–1958).

Gasser, Raymond, *Atlas of Human Embryos* (Hagerstown, MD: Harper and Row, 1975).

Gilbert, Steve, *Images of Embryos. An Illustrated Sourcebook in the History of Descriptive Embryology*, forthcoming.

Gonzalez-Crussi, Frank, *Suspended Animation: Six Essays on the Preservation of Bodily Parts* (New York: Harcourt Brace and Company, 1995).

Harvey, A. McGehee, "A new school of anatomy: The story of Franklin P. Mall, Florence R. Sabin and John B. MacCallum," *Johns Hopkins Medical Journal* 136 (1975), pp. 83–94.

Heard, Osborne O., "A photographic method of orienting serial sections for reconstruction," *Anatomical Record* 49 (1931), pp. 59–70.

Recollections–Reflections, Hayden Lake, Idaho. MS in the collections of the Human Developmental Anatomy Center, National Museum of Health and Medicine, Armed Forces Institute of Pathology, Washington, DC, 1979.

Hertig, Arthur W., *Human Trophoblast* (Springfield, IL: Charles C. Thomas, 1968).

His, Wilhelm, *Unsere Körperform und das physiologische Problem ihrer Enstehung.* (Vogel: Leipzig, 1874).

Hopwood, Nick, "Producing development: the anatomy of human embryos and the norms of Wilhelm His," *Bulletin of the History of Medicine* 74 (2000), pp. 29–79.

Embryos in Wax: Models from the Ziegler Studio (Cambridge: Whipple Museum of the History of Science, Cambridge University; and Bern: University of Bern, 2002).

Huber, G. C., "Franklin Paine Mall, 1862–1917," *Anatomical Record* 14 (1918), pp. 3–17.

Journal of Anatomy, "George Linius Streeter", *Journal of Anatomy* 83 (1949), pp. 51–2.

Keibel, F. and Elze, C., *Normentafeln zur Entwicklungsgeschichte des Menschen* (Jena: Fisher, 1908).

Keibel, Franz and Franklin P. Mall, *Handbook of Human Embryology* (Philadelphia: Lippincott, 1910 and 1912), 2 volumes.

Keller, Evelyn Fox, *Making Sense of Life: Explaining Biological Development with Models, Metaphors, and Machines* (Cambridge, MA: Harvard University Press, 2002).

Landecker, Hannah, "Technologies of living substance: tissue culture in twentieth century biomedicine," (Doctoral Thesis, Massachusetts Institute of Technology, 1999).

Lewis, Frederic T., "Charles Sedgwick Minot," in Howard A. Kelly and W. Burrage, *Dictionary of American Medical Biography* (Boston, MA: Milford House, 1928), pp. 849–51.

Lewis, W. H., "The use of guide planes and plaster of Paris for reconstructions from serial sections: some points of reconstruction, *Anatomical Record* 9 (1915), pp. 719–29.

McLaughlin, Loretta, *The Pill, John Rock, and the Church* (Boston, MA: Little, Brown, 1982).

Maienschein, Jane, "The origins of Entwicklungsmechanik," in Scott F. Gilbert (ed.), *A Conceptual History of Modern Embryology* (New York: Plenum Press, 1991), pp. 43–61.

Maienschein, Jane, *Transforming Traditions in American Biology, 1880–1915* (Baltimore, MD: Johns Hopkins University Press, 1991).

Mall, Franklin P., "A human embryo twenty-six days old," *Journal of Morphology* 5 (1891), pp. 459–80.

"Wilhelm His, his relation to institutions of learning," *American Journal of Anatomy* 4 (1905), pp. 139–61.

"A study of the causes underlying the origin of human monsters," *Journal of Morphology* 19 (1908), pp. 1–367.

"A plea for an institute of human embryology," *Journal of the American Medical Association* 60 (1913), pp. 1599–601.

Carnegie Institution of Washington *Year Book* 16 (1916–17).

Carnegie Institution of Washington *Year Book* 17 (1917–18).

Mall, Franklin Paine and Arthur William Meyer, "Studies on abortuses: a survey of pathologic ova in the Carnegie embryological collection," *Contributions to Embryology* 7 (Washington, DC: Carnegie Institution of Washington, 1921), pp. 1–364.

Medley, Nancy, "The uses of women: Franklin Mall and human embryology, 1900–1917." Paper presented to the Department of the History of Science, Medicine and Technology, Johns Hopkins University, 11 April 2002, pp. 1–33.

Meyer, Arthur William, *The Rise of Embryology* (Stanford, CA: Stanford University Press, 1939).

Morgan, Lynn M., "Materializing the fetal body, or, what are those corpses doing in biology's basement?," in Lynn M. Morgan and Meredith W. Michaels (eds.), *Fetal Subjects, Feminist Positions* (Philadelphia, PA: University of Pennsylvania Press, 1999), pp. 43–60.

"Embryo tales," in Sarah Franklin and Margaret Lock (eds.), *Remaking Life & Death: Towards an Anthropology of the Biosciences* (Santa Fe, CA: SAR Press, forthcoming).

"'Properly disposed of': a history of embryo disposal and the changing claims on fetal remains," *Medical Anthropology* 21 (2002), pp. 247–74.

Neverov, Oleg, "'His Majesty's cabinet' and Peter I's Kunstkammer," in O. Impey and A. MacGregor (eds.), *The Origins of Museums: The Cabinet of Curiosities in Sixteenth and Seventeenth Century Europe* (London: House of Stratus, 2001), pp. 71–79.

Newman, Karen, *Fetal Positions: Individualism, Science, Visuality* (Stanford, CA: Stanford University Press, 1996).

Nyhart, Lynn K., *Biology Takes Form: Animal Morphology and the German Universities, 1800–1900* (Chicago, IL: University of Chicago Press, 1995).

Oppenheimer, Jane M., *Essays in the History of Embryology and Biology* (Cambridge, MA: MIT Press, 1967).

O'Rahilly, Ronan, "One hundred years of human embryology," in Harold Kalter (ed.), *Issues and Reviews in Teratology*, vol. 4 (1988), pp. 81–128.

O'Rahilly, Ronan and Fabiola Muller, *Developmental Stages in Human Embryos Including a Revision of Streeter's "Horizons" and a Survey of the Carnegie Collection*, Publication 637 (Washington, DC: Carnegie Institution of Washington, 1987).

O'Rahilly, Ronan, F. Muller, G. M. Hutchins, and G. W. Moore, "Computer ranking of the sequence of appearance of 100 features of the brain and related structures in staged human embryos during the first 5 weeks of development," *American Journal of Anatomy*. 180 (1984), pp. 69–86.

Ramsey, Elizabeth M., "The Yale embryo," *Contributions to Embryology* 27 (Washington, DC: Carnegie Institution of Washington, 1938), pp. 85–93.

Sabin, Florence R., "Franklin Paine Mall: a review of his scientific achievement," *Science*, N.S., 47 (1918), pp. 254–61.

"Franklin Paine Mall," in Howard A. Kelly and W. Burrage, *Dictionary of American Medical Biography* (Boston, MA: Milford House, 1928), pp. 803–4.

Franklin Paine Mall, The Story of a Mind (Baltimore, MD: Johns Hopkins Press, 1934).

"Franklin Paine Mall, 1862–1917," *Biographical Memoirs of the National Academy of Sciences* 16 (1936), pp. 65–122.

Stafford, Barbara Maria, *Body Criticism: Imagining the Unseen in Enlightenment Art and Medicine* (Cambridge, MA: MIT Press, 1991).

Streeter, George, L. Carnegie Institution of Washington *Year Book* 20 (1920–1).

Carnegie Institution of Washington *Year Book* 22 (1922–3).

Carnegie Institution of Washington *Year Book* 26 (1926–7).

Carnegie Institution of Washington *Year Book* 30 (1930–1).

"Prenatal growth of the child," *News Service Bulletin* (Staff Edition) 4, Number 14 (Washington, DC: Carnegie Institution of Washington, 1937), pp. 126–32.

"Developmental horizons in human embryos. Description of age group XI, 13 to 20 somites, and age group XII, 21 to 29 somites," *Contributions to Embryology* 30 (1942), pp. 211–45.

"Developmental horizons in human embryos. Description of age group XIII, embryos about 4 or 5 millimeters long, and age group XIV, period of indentation of the lens vescicle," *Contributions to Embryology* 31 (1945), pp. 27–63.

(prepared for publication by C. H. Heuser and G. W. Corner), "Developmental Horizons in Human Embryos. Description of age groups XIX, XX, XXI, XXII, and XXIII, being the fifth issue of a survey of the Carnegie Collection," *Contributions to Embryology* 34 (1951), pp. 165–96.

"Developmental horizons in human embryos. Description of age groups XV, XVI, XVII, and XVIII, being the third issue of a survey of the Carnegie collection," *Contributions to Embryology* 32 (1948), pp. 133–203.

"Developmental horizons in human embryos (fourth issue). A review of the histiogenesis of cartilage and bone," *Contributions to Embryology* 33, (1949), pp. 149–69.

Trefil, James and Margaret Hindle Hazen, *Good Seeing, A Century of Science at the Carnegie Institution of Washington 1902–2002* (Washington, DC: Joseph Henry Press, 2002).

Tsiaras, Alexander, *From Conception to Birth, A Life Unfolds* (New York: Doubleday, 2002).

Welch, William H., *et al.*, "Memorial service in honor of Franklin P. Mall, Professor of Anatomy, Johns Hopkins University, 1893–1917," *Johns Hopkins Medical Bulletin* 29, (1918), pp. 109–23.

3

HOW RHESUS MONKEYS BECAME LABORATORY ANIMALS

ELIZABETH HANSON

The Rockefeller University, New York

Many accounts describe the 1950s, when rhesus macaques (*Macaca mulatta*) were used in studies to develop a vaccine against polio, as the time when these animals became widely accepted as models for humans in laboratory research. For that work, more than 200,000 monkeys were imported into the USA each year.[1] Rather than a new role for rhesus monkeys, however, the polio vaccine work represented research with macaques on an unprecedented scale. By the 1950s, the rhesus monkey was already well established in the laboratory, and its anatomy and physiology were known in detail. The foundational work had been accomplished with a breeding colony of macaques established in the 1920s at the Carnegie Institution of Washington (CIW) Department of Embryology. It was here, through a deliberate effort to invest in infrastructure and develop husbandry methods while investigating the primates' reproductive physiology, that the development of the rhesus macaque as a model organism for comparison to humans was pioneered. Through the work at the Department of Embryology, the place of the monkey in the laboratory was established, and the rhesus macaque became the non-human primate of choice for laboratory investigation.

The development of the CIW monkey colony is an example of a larger trend or transition in early-twentieth-century biology from studying specimens collected in the field to the controlled production and study of organisms in the laboratory. This transition accompanied a shift from descriptive morphology to experimental physiology. The new experimental approach required large quantities of live and fresh material, and often large quantities of the same species. Laboratories began to develop on-site animal colonies, and biological supply houses were organized to fill the demand for animals in research. The Wistar Institute, for example, began distributing its famous albino rat in 1906.[2]

The desire to maintain control of both quantity and quality of research material motivated the establishment of breeding colonies for research. This required a large investment in infrastructure, and was not easily achieved. "Control of the material" was an explicit reason for founding the CIW rhesus colony.[3] What the CIW researchers sought to control was the breeding of the animals, and in addition, control of the material to them meant knowledge of the animals' ages and of their medical histories over a period of years. Referring to the inbred strains of animals that were becoming available to researchers, the Embryology Department's Director George Streeter acknowledged in 1938 that "There can be no question as to the desirability of a 'standard monkey' which could be had in large quantity."[4] The cost of keeping monkeys, however, and their relatively long gestation period and singleton births, made them unsuitable for breeding in numbers large enough to obtain standardization in genetic terms. Even though it bred animals, the CIW colony was not self-sustaining; it produced disposable research material that had to be replaced from time to time by the purchase of animals from dealers.

Acquiring, keeping, and studying rhesus macaques required building and maintaining quarters for the animals, and the development of craft skills in keeping and breeding the animals and of techniques for working with very early embryos. This was a long-term project; twenty years passed between the first "pilot" macaque colony and the publication that achieved the CIW Department's primary goal, analysis of a catalog of monkey embryos of known age, from the one-cell stage on. During this period, the monkeys both served the immediate research needs of the Embryology Department – they were "the right organism for the job" of studying reproductive physiology – and proved useful for a much wider range of research projects.[5]

The monkey colony was the Department of Embryology's second major investment in infrastructural tools and techniques. The first was a collection of human embryos assembled by anatomist Franklin P. Mall. The Department had been founded in 1914 as a center for housing, expanding, and studying this collection, which became the largest collection of human embryos in the world. The embryos were preserved, sectioned, photographed, and reconstructed as wax and plaster models, in an effort to elucidate the stages of development of the human embryo. The monkey colony was the Department's next large-scale venture. In terms of the resources it required and its value as a tool, it was comparable to equipment like a big telescope for the CIW's astronomers (the CIW had supported the Mount Wilson Solar Observatory, founded near Pasadena in 1904).[6] It became part of an expanding infrastructure of materials needed to do science.

In retrospect, this investment can be seen as emblematic of the Department of Embryology's research style, part of a strategy that has enabled a small contingent of scientists to achieve success and influence far beyond what would be predicted by the Department's size (see Spradling, chapter 8, this volume). Launching new enterprises that later come into widespread use is one aspect of this CIW research style, as is the flexibility to abandon these enterprises as others take them up, and move on to new problems. Other characteristics of the Department's research style include long-term investment in research problems and the considered recruitment of creative, independent individuals to lead new endeavors.

All of these elements are evident in the history of the monkey colony. In the context of the Department of Embryology, the rhesus colony is a striking example of a novel enterprise that later entered standard practice, and it provides insight into the process by which an animal becomes widely used in laboratory research. As the CIW researchers gained knowledge about the monkey's anatomy and physiology, and learned how to keep a healthy colony of animals, the rhesus macaque was transformed from the sometime object of experiments on transmitting and treating infectious disease to the primate of choice for use in all sorts of laboratory investigation. The CIW colony became a sort of model colony, visited frequently by scientists from other institutions who wanted to learn how to keep the animals.

This chapter focuses on the period in the 1920s and 1930s when the CIW colony was founded and the role of the rhesus macaque in laboratory research was solidified at the CIW's Department of Embryology. The Department maintained a colony of rhesus macaques for studies in reproductive physiology until 1971. By that time, reproductive physiology no longer was central to the department's research program, and the monkey colony was transferred to the Rush Medical School in Chicago.[7]

Rhesus macaques and other primates had of course been used in laboratories before 1920. For the most part, the monkeys seem to have served as models in which to study human disease. Monkeys were infected with a microbe, and if they proved susceptible, the course of the disease could be followed, experimental treatments attempted, and perhaps, therapeutic serums for treating humans derived from the monkey's blood. Searching for an animal model in which to study polio, for example, Karl Landsteiner, working in Vienna, reported polio infection in a macaque in 1909.[8] If such animals did not die from their infections, they were often killed and dissected as part of research. The animals were a resource to be purchased and used up rather than maintained. The status of the rhesus monkey in the laboratory in 1920, as summed up by George Streeter, the Director of the Embryology Department, was that "the macaque has, to some extent, become a laboratory animal for medical investigations."[9]

That the macaque had been used in medical research was significant for its choice by CIW researchers as a model organism. The CIW Department grew out of a tradition of medical research, rather than embryology as it was studied at seaside research stations like the Marine Biological Laboratory at

Figure 3.1 George W. Corner was a young associate professor in the anatomy department of the Johns Hopkins School of Medicine when he started the first breeding colony of rhesus monkeys. (Courtesy The Alan Mason Chesney Medical Archives of The Johns Hopkins Medical Institutions.)

Woods Hole, Massachusetts, or the Naples Zoological Station. Researchers at the CIW Department focused on human embryology; they held little interest in the development of invertebrates. Specifically, the Department's program built on the interests of researchers at the Johns Hopkins Medical School. Anatomist Franklin P. Mall, founder of the Department, had been among the first faculty at Hopkins, and many of the Department's early researchers had worked under Mall at Hopkins. In addition, the CIW Embryology Department was physically located on the campus of the Johns Hopkins Medical School, near the Hopkins Anatomy Department. The two were closely allied in the first decades of the history of the Department of Embryology.

The story of the monkey colony begins, in fact, in the Hopkins Anatomy Department with an anatomist named George W. Corner (Fig. 3.1). Corner had earned his M.D. under Franklin P. Mall at the Johns Hopkins Medical School. Initially torn between a career as a practitioner – he intended to become a gynecologist – and a research career in reproductive physiology, he decided on the latter and joined the Hopkins Anatomy Department in 1913. There, under Mall's direction, he began a study of the corpus luteum, the small progesterone-secreting mass of tissue that forms after the rupture of an ovarian follicle in mammals. Mall thought a better understanding of the corpus luteum might provide a way to determine the age of early human embryos. Corner's research material was not human. For his work on the corpus luteum, he obtained the reproductive organs of swine from a local slaughterhouse. In 1914, he joined another Hopkins graduate, Herbert Evans, at the medical school at the University of California at Berkeley and continued his research on the mammalian reproductive cycle using swine as his model.

In 1919, firmly committed to a research career, Corner returned to the Hopkins anatomy department as associate professor. By this time, he wanted to study the reproductive cycle in an animal more comparable to humans. The sow has an estrus cycle; only primates menstruate. Practically nothing was known about the human menstrual cycle in general or the relationship between ovulation and menstruation more specifically. Corner's research questions could only be studied in a primate. In addition, he had become accustomed through his work on swine to having reliable access to large quantities of fresh research material. So Corner began to investigate the possibility of continuing his work in a colony of non-human primates.[10]

In August of 1920 George Corner visited the monkey house at the Bronx Zoo to interview the keepers about how to obtain, keep and breed the animals, with the idea of starting a laboratory colony of rhesus monkeys. Corner's research, and its later expansion at the CIW Department of Embryology, differed from studies of infectious disease in its focus on the anatomy and physiology of the monkey itself. This work required keeping the animals in good health over a period of years; despite the assurances of zookeepers, part of Corner's project was to prove that this was possible. Setting up the project required, among other things, building cages that provided access to the outdoors – the zoo men told him that "the secret of success lies in proper outdoor housing." The animals themselves were expensive compared with mice or even dogs, costing as much as $18 each.[11]

"It should be made clear that there is a large risk in any such project, wherever attempted," Corner wrote, "pneumonia, tuberculosis, malnutrition and all sorts of bad luck might wipe out the animals and force a new start at any time."[12] A new start would of course not only be expensive, but it would set back research. Estimating that it would cost about $600 in its first year, Corner began a project to study the relationship between ovulation and menstruation in the rhesus macaque in the spring of 1921. By April of that year he had obtained six macaques and housed them on the fifth-floor balcony of the Hunterian Laboratory at Hopkins. Corner later wrote that this was "the first monkey colony in the United States for long-term physiological experimentation."[13]

Corner's risk turned out to be well calculated. Within two years he had completed a neat project with clear and significant results. After observing the cycles of 11 female macaques and killing them at planned intervals in the cycle, he had mapped out for the first time the sequence of events – in particular the time of ovulation – of the menstrual cycle in primates.[14]

Researchers at the CIW Department of Embryology were familiar with Corner's work. Indeed, they knew Corner well. While Corner was in Berkeley, Mall founded the Department of Embryology, and others there including Warren Lewis knew Corner from his student days. The success of Corner's small monkey colony – a sort of pilot project – spurred the Department of Embryology to invest in the development of a larger breeding colony of rhesus macaques. These animals would serve as model organisms for comparison to human reproductive physiology and embryological development.

Corner left Baltimore in 1923 to head the anatomy department at the medical school being built at the University of Rochester, in New York. He started a new monkey colony there (all his original animals had been killed for research) and continued his work; he eventually discovered the hormone progesterone. (Much later in his career, in 1940, Corner returned to Baltimore to head the CIW Embryology Department.) Two years after Corner's departure, in 1925, George Streeter, the Director of the CIW Department of Embryology, proposed starting a much larger breeding colony of rhesus monkeys at the Department. Streeter was impressed with the scientific success of Corner's project – he had made pioneering contributions to understanding primate reproduction. But he seemed just as interested in Corner's success at showing the macaque could be kept as a source of research material. Streeter wrote:

> From the experiments conducted here by Dr. Corner, it has been shown that, with proper feeding and management, the common macaque monkey (Pithecus rhesus) thrives all the year around in the outdoor climate of Baltimore, even exhibiting, with normal periodicity, the sensitive phenomenon of menstruation. We know, also, that this species of monkey becomes pregnant and gives birth to normal young under the conditions of domestication that can be provided here; in zoological gardens as many as three generations have been recorded. Owing to their relatively small size, the question of housing and feeding these animals is not difficult. Furthermore, being a hardy species, it is the one most commonly sold by dealers and hence is obtainable at a reasonable price and in sufficient numbers for the purposes of a breeding colony.[15]

From the perspective of the twenty-first century, when macaques and many other primates are routinely maintained for laboratory research and in zoos, Streeter's emphasis on the observation that monkeys could survive in Baltimore seems curious and is easy to overlook. But at the beginning of the twentieth century, climate was considered an aspect of the environment crucial to the survival of monkeys and apes in captivity. It was an important element of Robert Yerkes' proposal in 1916 to establish a research station for the scientific study of monkeys and apes. "The institute should be located in a region whose climate is in high degree favorable to the life of a variety of lower primates and man," he wrote. Yerkes indicated too that some compromise would be needed between the "enervating tropical climate" in which non-human primates were presumed to thrive and "civilization," the habitat of scientists.[16] Even at zoos, there was much debate over whether primates were healthiest if kept indoors in heated cages. The ideas of the German animal dealer and showman Carl Hagenbeck were just beginning to change zookeeping practices around the turn of the twentieth century, and one of Hagenbeck's innovations in animal keeping had been to demonstrate that tropical species can live outdoors in a range of climates.[17] In addition, at zoos and elsewhere, nearly nothing was known about the natural behavior or diet of any of the monkeys or great apes. Successful acclimatization of these and other wild animals was not taken for granted.

Streeter's comment that monkeys were readily available also bears further elaboration. Rhesus monkeys were available to researchers because they were regularly imported to the USA as part of a thriving animal trade that supplied circuses and zoos at the turn of the twentieth century.[18] Other primate species also were imported, but were more rare and consequently more expensive. Ecology provides one explanation for the availability of rhesus macaques. They were imported mainly from areas of India where they were commensal with human populations, living near towns and in and around temples. Animal dealers did not have to go far to trap them.[19] Rhesus macaques also seemed to be hardier than other species, able to survive several months of travel in cages lashed to the decks of the ships that brought them to the USA. This may have been purely a numbers game – more rhesus macaques were captured and transported than other species, and therefore more survived the journey – rather than a reflection of their physiology. In any case, zookeepers knew that these animals survived well in captivity, and often bred. George Corner found this out when he talked to people at the Bronx Zoo as well as the zoos in Philadelphia and Washington, DC to find out what it would take to maintain a colony of the animals for laboratory research.

Breeding, or more properly, primate reproduction, was what researchers at the embryology department wanted to study. The collection of preserved human embryos yielded much valuable information, but it was not possible with this collection to study embryological development between the time of fertilization of the egg and the implantation of the embryo in the uterus about 14 days later. These embryos were so small, and so rarely recovered, that researchers simply did not have the material to analyze. With monkeys whose reproductive cycles were closely monitored it would be possible to recover and study these early stages of the embryo. Moreover, the ages of the embryos in the human collection had to be estimated. In contrast, monkey embryos could be collected at precise times after the animals mated. A complete series of monkey embryos of known age would make the human embryo collection more valuable, because comparative study would aid in the interpretation of the human material.

A collection of primate embryos parallel to the human embryo collection had in fact already been started. The CIW had funded an expedition to Nicaragua by the primatologist Adolph Schultz in the summer of 1924. Among other mammals, Shultz collected 191 monkey specimens for comparative anatomical study. Material saved from autopsies would help investigators back in Baltimore to study what he called the estrus cycle, and included a series of eight small embryos and "quite a number of larger fetuses."[20]

But for a steady supply of fresh research material, and for the controlled production of embryos of known age, a colony was needed. Comparative anatomy between macaques and humans, and macaques and other primates, was one job that Streeter had in mind for the monkey colony when he wrote his proposal to John C. Merriam, president of the CIW. A descriptive anatomy of the macaque and its developmental correlation with human development was needed for the continued use, and future success, of the animal as a tool for medical research. Streeter also outlined a research program focused on the rhesus colony. The colony would provide a unique opportunity to study prenatal physiology and behavior, and the development and function of various organs, in particular the brain. Studies on the physiology of pregnancy, including menstruation, ovulation, cyclic changes in the uterus, and the viability and functional activity of the egg and sperm would have practical applications. "All of these things must be determined if we are to understand sterility, abortion [miscarriage], normal and defective progress of pregnancy, and the associated menstrual ailments which the gynecologist is constantly called upon to treat," Streeter wrote. He continued, suggesting that Warren and Margaret Lewis might apply their tissue culture technique to studying the development of the fertilized egg "under the microscope." Furthermore, building on a line of inquiry that dated from the beginning of the department, physical anthropologists could make growth studies from the prenatal stages of macaque development, through infancy and the adult. To Streeter's mind there was no shortage of uses to which a healthy breeding colony could be put. In fact, Streeter wrote, "With a well-established colony of monkeys, one of the difficulties to be anticipated is to keep some investigators away from it, as those concerned with the medical and biological sciences will readily appreciate. It will be necessary to restrict its use to those working on problems definitely related to our own program."[21]

Setting up a colony required the kind of work that is essential to scientific practice but forgotten once it becomes routinized. Streeter went about this task in an ambitious, yet well-planned manner typical of new projects launched at the department. First, new facilities needed to be built in which to keep the animals, and Streeter proposed starting the colony with 100 monkeys. The cages were constructed on the roof of the New Hunterian building[22] (Fig. 3.2). For the first year of the monkey colony, Streeter

Figure 3.2 The CIW's Department of Embryology was housed in the New Hunterian building on the medical campus of The Johns Hopkins University. The fenced-in monkey runs on parts of the roof are visible in this undated photograph. (Courtesy The Alan Mason Chesney Medical Archives of The Johns Hopkins Medical Institutions.)

budgeted $10,000, which was 18 percent of the department's budget for 1926.[23]

By 1925, Streeter also had in mind a scientist who would get the colony up and running as well as conduct research. This was Carl G. Hartman, an associate professor of zoology at the University of Texas in Austin. Hartman had for several years studied the embryology and physiology of reproduction in the opossum. The opossum is a marsupial, so the fetuses develop in a pouch where they can be observed. Hartman's thorough study of nearly 1,000 female opossums from which he obtained several thousand eggs had yielded the stages of intra- and extra-uterine development of the animal. Hartman chose his model organism largely on the basis of convenience. When asked "why the opossum?" he replied, "The prairie and the creek bottoms were full of [them] and nobody had yet bothered to study the development of these, the most bizarre of all American mammals."[24]

At the University of Texas, Hartman was poorly paid (he sold real estate on the side), and he taught large classes of undergraduates in zoology. He had few resources for research.[25] He carried out all aspects of the opossum research himself, including trapping the animals, raising them in his home, operating on them, securing, fixing, sectioning, and staining the eggs, and photographing the sections.[26] He sometimes also sold the animals to other researchers interested in studying them. It was in this role, as a supplier of biological materials, that Lewis Weed, head of the Hopkins anatomy department, knew Hartman personally. Weed must have been familiar with Hartman's published work, but their correspondence in 1924 was about Hartman supplying Weed with pregnant opossums.[27]

By this time Hartman was already interested in extending his work to the physiology of primate reproduction, and he was planning an expedition to the Philippines to collect monkey embryos. For this project he had raised money from private donors and also had a grant from the National Research Council's Committee for Research in Problems of Sex.[28] In the 1920s, field scientists, commercial animal collectors and dealers, and zoos were still active participants in the network supplying biological materials to laboratories. Up to this point in his career Carl Hartman had worked in the field as much as the laboratory. But with his involvement with the CIW monkey colony he made a career change that paralleled the wider transition in biology from physiological research based in the field to the lab.

Although Hartman had worked hard to obtain the funding for his expedition, Streeter convinced him to give that up for the opportunity to study the monkeys in Baltimore, the "safe and sane" way, as he put it, in a breeding colony in a laboratory setting. The Department of Embryology was clearly making a big investment in this research program. Streeter assured Hartman that it was an "ambitious project" to be "executed in outstanding manner."[29] "The real decision," Streeter wrote to Hartman, "would be whether the character and conditions of the research program we are considering is the type you would like most to plunge into for the next 10 or 20 years."[30]

Hartman arrived in Baltimore late in 1925. He had to make sure the facilities for his monkeys were ready and acquire 100 or so of the animals (Fig. 3.3). Freed from teaching responsibilities, he began with his own research. By 1927, Hartman had for the first time accurately recorded the duration of pregnancy of the macaque. Among his first achievements was developing a method of palpating the monkey ovaries that allowed him to observe when ovulation had taken place. This required monkeys that were very tame. Hartman described the technique: "With the middle finger of one hand in the rectum, the free hand on the abdomen, it is possible to palpate the uterus and both ovaries of the female monkey with infinitely greater ease and definiteness than in human patients."[31] Even with this experience, Streeter wrote, "we still feel

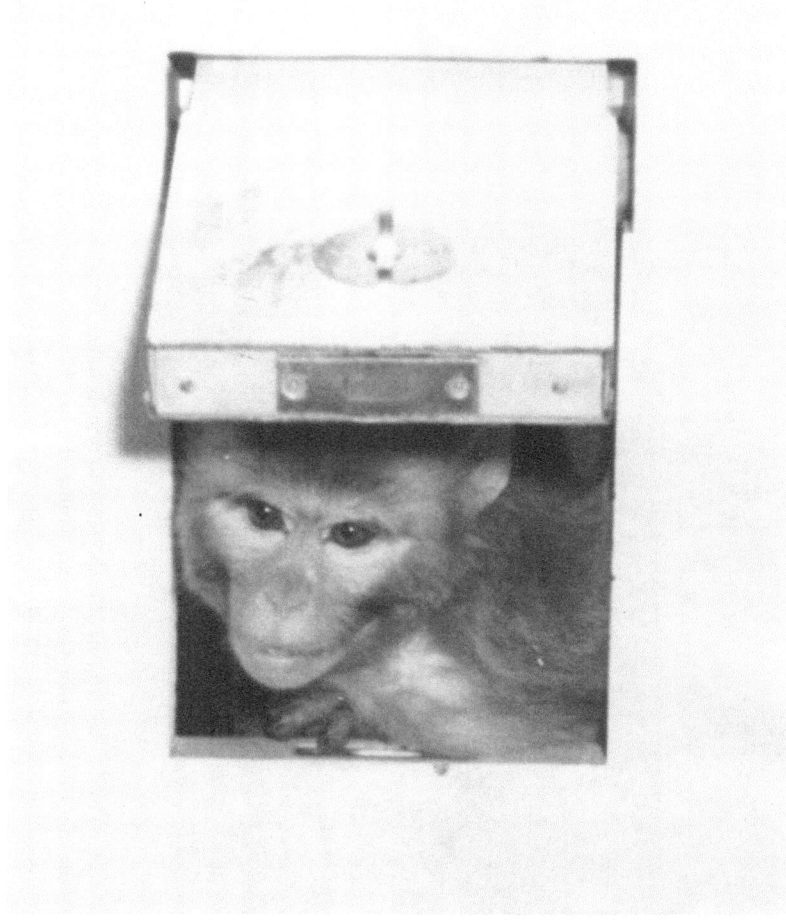

Figure 3.3 This four-year-old female rhesus monkey was born in the Carnegie colony.

that we are in the experimental stage as far as the handling of these animals is concerned."[32]

From the beginning, Hartman collaborated with many people, fulfilling Streeter's plan for the colony as a resource both locally at the embryology department and Johns Hopkins, and farther afield (for further details on Hartman's research, see Adele Clarke, chapter 4 this volume). He collaborated with George Corner (then in Rochester, NY) on studies of the corpus luteum, among other subjects. With his CIW colleague Warren Lewis, Hartman used the hanging-drop method of cell culture to study the timing of cell division of macaque zygotes to the 16-cell stage, and Lewis's time-lapse photography

Figure 3.4   Scientists working in the Department of Embryology in 1929 included, from left: Lewis, Streeter, Nordmann, Fauré-Fremiet, Grodzinskii, Hunter, Heuser, Hartman, and Tinklepaugh. (Courtesy The Alan Mason Chesney Medical Archives of The Johns Hopkins Medical Institutions.)

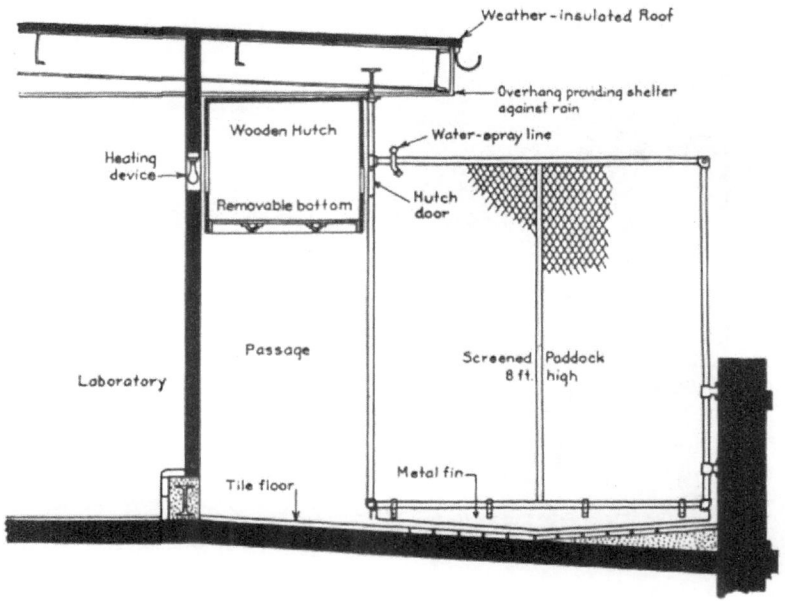

Fɪɢ. II. Section through a typical paddock (see text).

Figure 3.5   This sectional illustration shows some of the details considered in constructing quarters for the monkey colony. (Reproduced from Carl G. Hartman and William L. Straus, Jr., eds., *The Anatomy of the Rhesus Monkey* (Baltimore, MD: Williams & Wilkins, 1933), p. 363.)

technique was used to film this process (for further discussion, see Hannah Landecker, chapter 5 this volume). Hartman worked with others on problems of reproductive biology such as the development of the placenta. Researchers from outside institutions often came to work with the monkey colony as well, for example when a member of the Yerkes lab in New Haven came to study the births and newborn behavior of the macaques. By 1941, an important goal at the colony's founding had been achieved: through timed matings of the monkeys Hartman had produced the embryological material that Streeter and Chester Heuser analyzed and wrote up in *Developmental Horizons*, the first complete embryological series in a primate[33] (Fig. 3.4).

Not the least of Hartman's contributions to science was his accumulating expertise in keeping and breeding the macaques. Around 1930, based on what had been learned thus far, the monkey quarters were renovated to hold as many as 200 animals, and a steady stream of visitors came to the embryology department to tour the facilities.[34] Hartman published plans of his monkey quarters in CIW publications, and also in a volume that he edited in 1933, which was the first comprehensive anatomy of the rhesus macaque[35] (Fig. 3.5).

FIG. III. Sketch showing general character of paddocks and vestibule. Note arrangement of perches.

Figure 3.6   An animal keeper holds a net specially designed for catching monkeys in this illustration of the paddocks at the CIW Department of Embryology.
(Reproduced from Carl G. Hartman and William L. Straus, Jr., eds., *The Anatomy of the Rhesus Monkey* (Baltimore, MD: Williams & Wilkins, 1933), p. 363.)

He described in detail the arrangement of cages and paddocks, materials, perches, matters of heating and cooling, and so on, as well as what to feed the animals. The animals in the CIW colony were handled intensively, and attention to detail helped accomplish this safely. Hartman offered wry advice in his section on catching the animals, writing that "There should always be more than one door between the monkeys and the freedom of the outer world."[36] George Corner also found humor in his first attempts at keeping monkeys, writing that "When the medical students were at work in the dissecting rooms I was in full view catching the monkeys and often got a jolly round of applause from my pupils."[37] But Hartman endeavored to take the chance, if perhaps the entertainment value, out of this work. He designed an iron-framed net for catching the monkeys, with "features found by trial and error to be efficient." Hartman wrote, for example, "The top margin is straight so as to prevent the monkey's dodging out from under the corners"[38] (Fig. 3.6). As Hartman became the authority on rhesus monkey colonies, others, for example from the Rockefeller Institute, changed their animal-keeping practices based on his example.[39] The same year as his cage designs were first published, 1932, Hartman also summarized his scientific work to date in a pioneering monograph on menstruation and pregnancy in the macaque.[40]

By the 1930s it could be said that the rhesus macaque had become a laboratory animal. Streeter credited Hartman with making this possible, writing in 1941 that "step by step, he did succeed in converting this small, rugged, easily obtained, and relatively inexpensive primate into a laboratory animal."[41] When, in 1940, the primatologist Clarence Ray Carpenter proposed establishing a monkey colony to supply laboratories, George Streeter interpreted this effort as "an expression of the importance that experimentation on monkeys has now attained."[42] The numbers of macaques imported into the USA, mainly for scientific and medical research, also attest to the animal's increasing importance in the laboratory, reaching more than 12,000 monkeys per year by the late 1930s.[43]

Hartman himself observed that, "the rhesus monkey is the one which has established itself among us as a laboratory animal," adding, "I doubt whether the popularity of this species is due to superior qualities over others. It is probably availability that is responsible."[44] But Hartman, for his part, was unduly modest. Availability is not enough to make an animal widely used in laboratory investigation. Hartman played a large role in the transformation of the rhesus macaque into a laboratory animal, and others acknowledged his role. As Kohler has shown for *Drosophila*, and others have demonstrated for other organisms, accumulated data and research experience with a particular organism help sustain its use in the laboratory.[45] The story of the CIW macaque colony captures a period when many organisms were being acclimatized, bred, and standardized for laboratory investigation, a time of transition in biology from studying specimens collected in the field to controlled production and study of organisms in the laboratory. Hartman's career bridged this transition. His earlier experience with all stages of study of the opossum, beginning with capturing and keeping the animal, prepared him ideally for the painstaking day-to-day management of a macaque colony. Hartman's work at the CIW Department of Embryology to standardize the keeping and breeding of the rhesus monkey, and to understand its anatomy and physiology, made the macaque valuable to others and solidified its place as "the monkey version of a laboratory rat."[46]

Acknowledgements

I am grateful to all of the authors in this volume, but especially to Adele E. Clarke, for their help in developing this article. My thanks also go to the archivists at the Alan Mason Chesney Medical Archives at The Johns Hopkins Medical Institutions, the Rockefeller Archive Center, and especially John Strom and Tina McDowell at the CIW. Without the vision and persistence of Maxine Singer, former President of the CIW, this volume would not have come to fruition. I would also like to acknowledge Margaret Hazen, who was integral to this project from its beginnings, and who generously shared her broad knowledge of the CIW and its history. An earlier version of this chapter was presented at the History of Science Society annual meeting in 2002, in a session with other chapters in this volume, and benefited from comments offered there.

Notes

1. Deborah Blum, *The Monkey Wars* (New York: Oxford University Press, 1994), pp. 45–6; Andrew N. Rowan, *Of Mice, Models, & Men: A Critical Evaluation of Animal Research* (Albany, NY: State University of New York Press, 1984), p. 110.

2. Adele Clarke, "Research materials and reproductive science in the United States, 1910–1940," in Susan Leigh Star (ed.), *Ecologies of Knowledge: Work and Politics in Science and Technology* (Albany, NY: State University of New York Press, 1995), pp. 183–225; Bonnie Tocher Clause, "The Wistar rat as a right choice: establishing mammalian standards and the ideal of a standardized mammal," *Journal of the History of Biology* 26 (1993), pp. 329–49.

3. Chester H. Heuser and George L. Streeter, "Development of the macaque embryo," *Contributions to Embryology*, no. 181 (Washington, DC: Carnegie Institution of Washington, 1941), p. 17.

4. Streeter to Weed, 12 May 1938, Alan Mason Chesney Medical Archives, The Johns Hopkins Medical Institutions, record group CIWDE, box 45, f. "Weed, Lewis, 1919–1937."

5. "The right organism for the job," Special Section, *Journal of the History of Biology* 26 (1993), pp. 235–367; see also, Adele E. Clarke and Joan H. Fujimura (eds.), *The Right Tools for the Job: At Work in Twentieth Century Life Sciences* (Princeton, NJ: Princeton University Press, 1992).

6. James Trefil and Margaret Hindle Hazen, *Good Seeing: A Century of Science at the Carnegie Institution of Washington, 1902–2002* (Washington, DC: Joseph Henry Press, 2002), p. 105.

7. Carnegie Institution of Washington *Year Book* 70 (1970–1), p. 84.

8. K. Landsteiner and E. Popper, "Uebertragung der Poliomyelitis acuta auf Affen," *Zeitschrift für Immunitatsforschung und Experimentelle Therapie. T.I. Originale* 2 (1909), pp. 377–90. In the 1930s, Landsteiner's work with rhesus macaques led to his discovery of the Rh factor in blood.

9. George L. Streeter, "Memorandum: of the proposed breeding of rhesus monkeys at the department of embryology," 20 April 1925, CIW Archives, Embryology, Dir. 1913–35, f. 1.

10. Adele Clarke, "Research materials and reproductive science," pp. 203–6.

11. George Corner to Lewis Weed, 10 September 1920, Alan Mason Chesney Archives, The Johns Hopkins Medical Institutions, Lewis Weed papers, box 7, folder Corner 1918–24.

12. Ibid.

13. George W. Corner, *The Seven Ages of a Medical Scientist: An Autobiography* (Philadelphia, PA: University of Pennsylvania Press, 1981), p. 164.

14. Ibid., pp. 163–6.
15. Streeter "Memorandum," 20 April 1925.
16. Robert M. Yerkes, "Provision for the study of monkeys and apes," *Science* 43 (1916), p. 231.
17. Carl Hagenbeck, *Beasts and Men*, abridged translation by Hugh S. R. Eliot and A. G. Thacker (London: Longmans, Green, and Co., 1909), p. 202; see also Nigel Rothfels, *Savages and Beasts: The Birth of the Modern Zoo* (Baltimore, MD: Johns Hopkins University Press, 2002).
18. On animal collectors and dealers, see for example Elizabeth Hanson, *Animal Attractions: Nature on Display in American Zoos* (Princeton, NJ: Princeton University Press, 2002).
19. On the ecology and behavior of rhesus monkeys in India, see Charles H. Southwick, Mirza Azhar Beg, and M. Rafiq Siddiqi, "Rhesus monkeys in North India," in Irven DeVore (ed.), *Primate Behavior: Field Studies of Monkeys and Apes* (New York: Holt, Rinehart and Winston, 1965), pp. 111–59.
20. Adolph H. Schultz to John C. Merriam, 19 September 1924, CIW Archives, Embryology, Misc. 1919–46, f. 4.
21. Streeter "Memorandum," 20 April 1925.
22. Weed to Streeter, 25 April 1925, Alan Mason Chesney Archives, The Johns Hopkins Medical Institutions, CIWDE box 45, f. 21.
23. See Appropriations, in Alan Mason Chesney Archives, The Johns Hopkins Medical Institutions, RG 3, Ser. 1, Financial Records, box 1, f. 2.
24. Rudolph P. Vollman, "Carl Hartman's contributions to the physiology of reproduction," in Rudolph P. Vollman (ed.), *Fifty Years of Research on Mammalian Reproduction: A Bibliography of the Scientific Publications of Carl G. Hartman* (Washington, DC. USDHEW, Public Health Service Publication no. 1281, 1965).
25. Real estate, Carl Hartman to George L. Streeter, 28 April 1925, Alan Mason Chesney Archives, The Johns Hopkins Medical Institutions, CIWDE box 21, f. Hartman (1); departmental responsibilities, box 21, f. 31.
26. Vollman, "Carl Hartman's contributions."
27. Weed to Hartman, 23 January 1924 and Hartman to Weed, 26 January 1924, Alan Mason Chesney Archives, The Johns Hopkins Medical Institutions, Lewis Weed papers, box 16, f. Hartman, C., 1921–52.
28. On Hartman's planned Philippines expedition, see correspondence between Hartman and Streeter, Alan Mason Chesney Archives, The Johns Hopkins Medical Institutions Chesney Medical Archives, CIWDE box 21, f. Hartman (1).
29. Streeter to Hartman, 22 May 1925, Alan Mason Chesney Archives, The Johns Hopkins Medical Institutions, CIWDE box 21, f. Hartman (1).
30. Streeter to Hartman, 7 May 1925, Alan Mason Chesney Archives, The Johns Hopkins Medical Institutions, CIWDE, box 21, f. Hartman (1).
31. Carl G. Hartman, "Bimanual rectal palpation as applied to the female rhesus monkey," *Anatomical Record* 45 (1930), p. 263.
32. Streeter to Robert M. Yerkes, 29 October 1927, Alan Mason Chesney Archives, The Johns Hopkins Medical Institutions, CIWDE, box 48, f. "Yerkes."
33. Streeter to Merriam, 21 March 1929, CIW Archives, Embryology – Director, f. 2.
34. See for example, Lewis Weed to Simon Flexner, 22 October 1932, Rockefeller Archive Center, RU, RG 301.2, box 5, f.3.
35. Carl G. Hartman and William L. Straus, Jr. (eds.), *The Anatomy of the Rhesus Monkey* (Baltimore, MD: Williams and Wilkins, 1933), pp. 361–71.

36. Carl G. Hartman, "Appendix: housing and care," in Carl G. Hartman and William L. Straus, Jr. (eds.), *The Anatomy of the Rhesus Monkey* (Baltimore, MD: Williams & Wilkins, 1933), p. 367.
37. Corner, *Seven Ages*, p. 164.
38. Hartman, "Appendix: housing and care," p. 367.
39. On the Rockefeller Institute, see Waldo Flinn to Simon Flexner, 12 November 1932, Rockefeller Archive Center, RU, RG 301.2, box 5, f. 3.
40. Carl G. Hartman, "Studies in the reproduction of the monkey, *Macacus (Pithecus) rhesus* with special reference to menstruation and pregnancy," *Contributions to Embryology*, no. 134 (Washington, DC: Carnegie Institution of Washington, 1932), pp. 1–161.
41. Chester H. Heuser and George L. Streeter, "Development of the macaque embryo," p. 17.
42. Richard G. Rawlins and Matt J. Kessler, "The History of the Cayo Santiago colony," in *The Cayo Santiago Macaques: History, Behavior and Biology* (Albany, NY: State University of New York Press, 1986), pp. 13–45; quotation is from Streeter to Weed, 12 May 1938, Alan Mason Chesney Medical Archives, The Johns Hopkins Medical Institutions, CIWDE, box 45, f. "Weed, Lewis, 1919–1937."
43. C. R. Carpenter, "Rhesus monkeys (*Macaca mulatta*) for American laboratories," *Science* 92 (1940), p. 284.
44. Carl G. Hartman, "The mating of mammals," in *Animal Colony Maintenance*, *Annals of the New York Academy of Sciences*, vol. 46 (New York: New York Academy of Sciences, 1945), p. 39.
45. Robert E. Kohler, *Lords of the Fly* (University of Chicago Press, 1994); "The right organism for the job," pp. 235–367.
46. Deborah Blum, *The Monkey Wars*, p. 246.

Bibliography

Blum, Deborah, *The Monkey Wars* (New York: Oxford University Press, 1994).
Carnegie Institution of Washington *Year Book* 70 (1970–1).
Carpenter, C. R., "Rhesus monkeys (*Macaca mulatta*) for American laboratories," *Science* 92 (1940), p. 284.
Clarke, Adele, "Research materials and reproductive science in the United States, 1910–1940," in Susan Leigh Star (ed.), *Ecologies of Knowledge: Work and Politics in Science and Technology* (Albany, NY: State University of New York Press, 1995), pp. 183–225.
Clarke, Adele E. and Joan H. Fujimura (eds.), *The Right Tools for the Job: At Work in Twentieth Century Life Sciences* (Princeton, NJ: Princeton University Press, 1992).
Clause, Bonnie Tocher, "The Wistar rat as a right choice: establishing mammalian standards and the ideal of a standardized mammal," *Journal of the History of Biology* 26 (1993), pp. 329–49.
Corner, George W., *The Seven Ages of a Medical Scientist: An Autobiography* (Philadelphia, PA: University of Pennsylvania Press, 1981).
Hanson, Elizabeth, *Animal Attractions: Nature on Display in American Zoos* (Princeton, NJ: Princeton University Press, 2002).
Hagenbeck, Carl, *Beasts and Men*, abridged translation by Hugh S. R. Eliot and A. G. Thacker (London: Longmans, Green, and Co., 1909).

Hartman, Carl G., "Bimanual rectal palpation as applied to the female rhesus monkey," *Anatomical Record* 45 (1930), p. 263.

"Studies in the reproduction of the monkey, *Macacus* (*Pithecus*) *rhesus* with special reference to menstruation and pregnancy," *Contributions to Embryology*, no. 134 (Washington, DC: Carnegie Institution of Washington, 1932), pp. 1–161.

"Appendix: housing and care," in Carl G. Hartman and William L. Straus, Jr. (eds.), *The Anatomy of the Rhesus Monkey* (Baltimore, MD: Williams & Wilkins, 1933), p. 367.

"The Mating of Mammals," in *Animal Colony Maintenance, Annals of the New York Academy of Sciences*, vol. 46 (New York: New York Academy of Sciences 1945), p. 39.

Hartman, Carl G. and William L. Straus, Jr. (eds.), *The Anatomy of the Rhesus Monkey* (Baltimore, MD: Williams and Wilkins, 1933), pp. 361–71.

Heuser, Chester H. and George L. Streeter, "Development of the macaque embryo," *Contributions to Embryology*, no. 181 (Washington, DC: Carnegie Institution of Washington, 1941), p. 17.

Kohler, Robert E., *Lords of the Fly* (Chicago, IL: University of Chicago Press, 1994).

Landsteiner, K. and E. Popper, "Uebertragung der Poliomyelitis acuta auf Affen," *Zeitschrift fur Immunitatsforschung und Experimentelle Therapie. T.I. Originale* 2 (1909), pp. 377–90.

Rawlins, Richard G. and Matt J. Kessler, "The history of the Cayo Santiago colony," in *The Cayo Santiago Macaques: History, Behavior and Biology* (Albany, NY: State University of New York Press, 1986), pp. 13–45.

"The right organism for the job," Special Section, *Journal of the History of Biology* 26 (1993), pp. 235–367.

Rothfels, Nigel, *Savages and Beasts: The Birth of the Modern Zoo* (Baltimore, MD: Johns Hopkins University Press, 2002).

Rowan, Andrew N., *Of Mice, Models, & Men: A Critical Evaluation of Animal Research* (Albany, NY: State University of New York Press, 1984).

Southwick, Charles H., Mirza Azhar Beg, and M. Rafiq Siddiqi, "Rhesus monkeys in North India," in Irven DeVore (ed.), *Primate Behavior: Field Studies of Monkeys and Apes* (New York: Holt, Rinehart and Winston, 1965), pp. 111–59.

Trefil, James and Margaret Hindle Hazen, *Good Seeing: A Century of Science at the Carnegie Institution of Washington, 1902–2002* (Washington, DC: Joseph Henry Press, 2002).

Vollman, Rudolph P., "Carl Hartman's contributions to the physiology of reproduction," in Rudolph P. Vollman (ed.), *Fifty Years of Research on Mammalian Reproduction: A Bibliography of the Scientific Publications of Carl G. Hartman* (Washington, DC: USDHEW, Public Health Service Publication no. 1281, 1965).

Yerkes, Robert M., "Provision for the study of monkeys and apes," *Science* 43 (1916), p. 231.

4

REPRODUCTIVE SCIENCE, 1913–1971

ADELE E. CLARKE

Department of Social and Behavioral Sciences, University of California

From its inception, the Carnegie Institution of Washington (CIW) Department of Embryology served as a Mecca for scientists from multiple disciplines and heterogeneous backgrounds who were interested in the reproductive anatomy and physiology of primates (including comparative anatomy) as well as in their embryonic development. In the early decades of the twentieth century, there were very few centers of reproductive research anywhere in the world. The scope and depth of reproductive studies undertaken at the CIW Department of Embryology could arguably be said to justify calling it the major medically oriented center in the USA during this period. This chapter examines the full range of research on reproductive phenomena pursued in the Department from its inception in 1913 to 1971, when this emphasis ceased.[1]

There are several key themes:

- First, like genetics in the USA, the reproductive sciences emerged from embryology. This was true in both biology (e.g., at the University of Chicago) and in medicine (e.g., at the CIW Department of Embryology). In contrast, in Britain, initial emphasis was on endocrinology, while in Germany reproductive research emerged largely out of gynecology. The Department is thus exemplary of the American case.[2]

- Second, the Department investigators who pursued reproductive phenomena composed an ongoing scientific community of intense collaborations, often over decades. These occurred both with each other and with regular and occasional visitors to the lab. Fostering sustained collaborative work has been an ongoing philosophy of the Department clearly manifest among its reproductive scientists.[3]

- Third, the work was organized around excellent access to particular research materials – from the local abattoir, the emergent primate colony, and, of course, the embryo collection. The CIW and the Department seriously invested in modern scientific research infrastructure by establishing access to these important materials just as the shift from morphological to experimental physiological approaches to research was manifest. This was a most innovative contribution to scientific research in its day. The investigators who developed and used the primate colony called themselves "the monkey fraternity."[4]

• Fourth, the CIW investigators and their collaborators pursued reproductive research despite its contested legitimacy. During the era of the formation of the US reproductive sciences in the decades after 1910, it was a maverick act – a radical social, political, cultural and moral act – to pursue any reproductive topics even in scientific venues because they were so deeply controversial. Franklin Paine Mall, the first Director of the Department, and others countered such challenges with assertions that basic research would lead to valuable clinical innovations – and it did.[5]

Let me begin by quickly painting the broad landscape of the development of the US reproductive sciences in biology, medicine, and agriculture into which this portrait of one center of medical reproductive research fits.[6] During the era of emergence *c.* 1910–25, the problem structures of reproductive research across the three professions varied. Biologists tended to focus on analytic problems such as sex determination, differentiation, and fertilization. Medical reproductive scientists tended to focus on the reproductive system in humans. Agricultural scientists tended to focus on the reproductive system in particularly profitable and manipulable domestic organisms. Across all three professional domains, the major problems addressed during the 1910–25 era were fertilization, sex differentiation, the estrus and menstrual cycles, ovarian function, and the corpus luteum. Both medical and agricultural researchers focused on the estrus cycle. There were also some beginnings of reproductive endocrinology as well, including the discovery of estrogens by the mid-1920s.

From *c.* 1925 to 1940, the reproductive sciences coalesced around endocrinology such that this was known as the "heroic age of reproductive endocrinology" or the "endocrinological gold rush." During this period, the chief naturally occurring estrogens, androgens, and progesterone were isolated and characterized, and the anterior pituitary, placental, and endometrial gonadotropins were also discovered. These accomplishments were marked by the publication of *Sex and Internal Secretions*, which fast became the US "bible" of the reproductive sciences. Two and a half inches (6.5 cm) thick, this tome marked the shift of preeminence in the reproductive sciences from Britain to the USA, where it essentially remained for half a century if not longer. The two main institutional homes of the scientists involved in producing this text were the University of Chicago and the CIW, each with four contributors.[7]

The US reproductive sciences have had an odd funding career. Despite serious legitimacy problems as a line of scientific work, they were quite successful in obtaining private research funding prior to World War II, largely from highly prestigious sources well within the mainstream of the biomedical research community. The stature of the sources, including three National Research Council committees, the Rockefeller, Macy, and Markle Foundations, and the CIW, was particularly significant. Not until the 1960s was there any federal funding of such research and, since then, it has been episodic and highly contested, while other private foundations have also stepped in. Thus, the CIW Department of Embryology was clearly a key player in the development and coalescence of the US reproductive sciences.[8]

Creating a major center of reproductive research in medicine

Medical interest in reproduction and maternal health had been relatively slow to develop and, when it did, it reflected broader efforts to expand medical jurisdictions and reproductive specialties. At the turn of the twentieth century, medical reformers wanted science to reign in all segments of the profession. One of the major reformers in obstetrics and gynecology was J. Whitridge Williams, head of the Johns Hopkins School of Medicine's Obstetrics Department. He started campaigning for more anatomical and pathological studies of the female generative tract in the 1890s.[9] In his 1914 Presidential Address to the American Gynecological Society, Williams delivered a "scathing reproach" to his colleagues because, in reviewing articles in the Society's *Transactions*, he had failed to find a single "fundamental" contribution to obstetrics. There was "an entire absence of reference to the biochemical aspects of pregnancy;" obstetricians and gynecologists placed "technical virtuosity" – largely in surgery – above serious attempts "to extend the limits of knowledge."[10] Williams' students, among others, remained less than enthusiastic about basic research on both educational and economic grounds; they were ill-prepared for it and it did not pay well. In 1925, Williams lamented that in terms of obstetrics and gynecology, most American medical schools remained half a century behind those in Germany.[11]

The call to research on reproduction was amplified by Williams' colleague at Johns Hopkins, Franklin Paine Mall, head of the Anatomy Department. Mall had been educated in Germany, working both with the embryologist Wilhelm His and the physiologist Karl Ludwig at Leipzig.[12] In 1913, Mall offered his student George Washington Corner an assistantship for teaching and research in anatomy just as Corner was also offered a prestigious internship in gynecology under Howard Kelly at the Hopkins Hospital. To convince Corner to come to anatomy, Mall argued for a "sounder scientific base in the clinical branches of medicine," and told Corner that he could "do more for the future of gynecology by basic research on embryology and the physiology of the reproductive system than I could if I merely followed . . . the static program of the distinguished gynecologists." Corner became a convert:

> During my year in Anatomy (1913–1914) I . . . began work on a problem which Mall himself now set for me: I was to study the cytological development of the corpus luteum with the hope of learning to use it as a measure of the age of early embryos . . . It was the beginning of my lifelong program of research on the histology and physiology of the reproductive system.[13]

One year later in 1914, Corner recalled:

> . . . I had a much better idea of the normal female reproductive cycle and the concomitant changes in the ovaries and cervix than did the average intern; indeed I may say that I knew more about the physiology of the reproductive organs than did the chiefs of the service, Howard Kelly and Thomas S. Cullen, world renowned leaders as they were in pelvic surgery and pathology. Gynecologists' . . . efforts to treat the functional disorders of menstruation and sterility were mere puttering, scarcely advanced beyond the procedures of the Hippocratic era. How could we hope for anything better when we simply did not understand the human cycle?

Corner devoted most of his professional life to understanding the female cycle. A major history of reproductive research actually periodizes reproductive science as "Before Corner" and "After Corner." [14] He was initially at the University of California at Berkeley, then served as founding chair of anatomy at the University of Rochester Medical School. Throughout his career, Corner was a regular visitor to the Department and frequent collaborator. In 1940, he became Director of the Department upon George Streeter's death, continuing until 1955.[15]

Both the primary funding sources for research in the life sciences at this time, the Rockefeller and Carnegie Foundations, strongly promoted the development of scientific entrepreneurs and a "team" or "center" approach.[16] Mall then became such a scientific entrepreneur, securing the CIW's support for a Department of Embryology, established at Johns Hopkins Medical School in 1913/14. It became the longest-lived example of direct foundation funding of a center of reproductive research (1913–71).[17] An institution at the heart of reproductive morphological, physiological and endocrinological research, the Department also published a major journal in the field, *Carnegie Contributions to Embryology* (1915–66). As a later Director, James Ebert, noted,

> The Department was for five decades the world's leading center for the study of the human embryo. It pioneered in the development of primates for research, having the earliest successful American monkey-breeding colony. Using these animals, large strides were made toward understanding menstruation and cyclic changes in the ovaries and uterus, laying much of the groundwork for recent advances in family planning.[18]

Many if not most of the medically oriented reproductive scientists from the USA and abroad working in the decades up to *c.* 1955 (the end of the Corner directorship) spent some segments of their careers there or worked closely with its faculty. Corner wrote: "During the sixteen years of my administration [1940–55], thirty-five men and women were listed formally in the *Year Books* . . . as fellows of the Guggenheim . . . or Rockefeller Foundation or one or another foreign organization in South America, Europe and Asia."[19]

The establishment of the American Board of Obstetrics and Gynecology in 1930 instituted needed reforms. Subsequently, obstetrics and gynecology merged more thoroughly, fusing women's reproductive health care under one specialty and further segmenting that specialty from both general practice and general surgery. In the USA in 1923, there were 696 full-time obstetrics/gynecology specialists; by 1949 there were 5074.[20] The shift to scientific medicine in gynecology and obstetrics was largely from surgical anatomy to reproductive physiology. It involved the development of functional (physiological) understandings of reproductive systems and processes to increase potential non-surgical therapeutics. Ironically, these alternatives were developed largely by anatomists and some physiologists, a considerable number of whom worked at the CIW Department of Embryology.[21]

A distinctive style of anatomy and physiology

Next I turn to the reproductive research done at the CIW Department of Embryology. While many of the investigators also worked with the embryo collection, I do not take up these projects here. I begin by noting that there was a distinctive style of anatomy in the CIW Department and at Hopkins more broadly. This style was not unlike the broader style of biological research at the University of Chicago led by Charles Otis Whitman, where Mall had spent a year and also where many regular visitors to the Department were based (including Bartelmez and Markee). Students at Hopkins had long been exposed to a unique combination of biological faculty: H. Newell Martin offering physiology in a medical framework (prior to the development of a medical school) and W. K. Brooks offering evolutionary morphology and comparative anatomy.[22] As Pauly put it, ". . . their students began to take seriously their pursuit of biology – seen as an intermingling of animal (largely invertebrate) physiology and morphology."[23]

Florence Sabin, a mainstay of the Hopkins anatomy faculty, chronicled the breadth at Chicago and Hopkins among anatomy faculty. Such faculty in numerous instances led the way to innovations in medical science by seeing themselves as equally close to zoologists involved in the new experimental biology as they were to medicine. Mall was likely the definitive contributor to this breadth as he was known in the USA for his zoological leanings and basic research advocacy. Corner, a Hopkins alumnus of both the undergraduate and Medical School, also noted a distinctive breadth on the medical side in anatomy.[24]

But the style in the Department was more than breadth, as biologist I. D. Raacke has described:

> Actually, the disciplinary labels are modern ones, for all considered themselves anatomists. As George Washington Corner . . . said only a few years ago: "I never did and still do not see any reason to call myself anything more than an anatomist" – even though his later work on the estrus cycle of the sow and the isolation of progesterone would now be described as physiological or even biochemical. In their tradition anatomy was not only the grandest of the biological sciences but the only real one, since *it rejected as artificial any separation between structure and function.* Anatomy, therefore, comprised not only gross and microscopic anatomy (histology) and embryology (the development of structure), but also physiology and all its subdivisions . . . [These are] the conceptual traditions of the integrated structure–function school of anatomy . . ."[25]

Thus was the Hopkins/CIW Department of Embryology tradition established.

This is also a juncture where the commitment of the Department to comparative anatomy should be noted. This was accomplished by hiring, in 1916, a physical anthropologist on his way to becoming a primatologist, Adolph H. Schultz. Schultz was initially responsible for producing an anthropometric record of the human embryo collection. But by 1921, he was also studying non-human primates, including sponsoring and going on collecting expeditions. When Schultz moved over to the Hopkins Department of Anatomy in 1925, he then maintained the colony's growth records and was given monkey material for his comparative work.[26]

Carl Hartman, who initially taught biology in Texas, was recruited into the Department by George Streeter in 1925, explicitly to develop the primate colony. After thirteen years of research in Texas, Carl Hartman had published extensively on the female cycle in the opossum. He then decided to study primates, and had planned a collecting trip to the Philippines. Instead, George Streeter intervened, inviting Hartman "to join his laboratory and study the monkey embryology the safe and sane way, bringing the monkey to the laboratory" at the CIW Department of Embryology.[27] Hartman then spent most of his career in the Department (see also Hanson, chapter 3 this volume). Part of the first generation of US reproductive scientists, Hartman too was committed to a broader view and was quite insistent that he be referred to as a reproductive physiologist rather than endocrinologist to reflect his commitment throughout his career.[28] So it was within such a non-reductionist view of anatomy and physiology that Department research on reproduction proceeded.

Figure 4.1 George W. Corner.

The reproductive cycle in human and non-human primates

By the early 1930s, some of the research done by the first generation of reproductive scientists produced the first seemingly accurate understanding of the female reproductive cycle. This was fundamental to the later development of most subsequent means of contraception from rhythm and other "natural" modes to the Pill, injectables, and so-called vaccines, but more immediately to physiological diagnostics and functional therapeutics.

Corner, Hartman and the female cycle

The maverick reproductive scientists most responsible for producing this new knowledge of the female cycle were founders and giants of the emergent discipline. George Corner (Fig. 4.1), a physician and fledgling reproductive scientist, had begun pursuing the cycle in 1914, at Mall's explicit urging noted

Figure 4.2   Carl G. Hartman.

above. This work culminated in a series of experiments on the rhesus monkey, *Macaca*, at Johns Hopkins Medical School to determine the parameters of the menstrual cycle. This work led, by 1929, to Corner's understanding of the action of progesterone, an essential hormonal actor in the menstrual cycle.[29] While he was later Director of the Department, his early research was not officially under its auspices although he was one of Mall's main protégés.

Carl Hartman (Fig. 4.2) then pioneered a mode of rectal palpation of the ovaries that could be used daily to estimate the time of ovulation and its relationship to phases of the menstrual cycle.[30] Through timed matings, he provided needed research materials for both embryological and reproductive work. With George Corner, he worked on the histogenesis of the corpus luteum. Hartman himself showed that the frequency of anovulatory cycles explains relative sterility in adolescence (discussed below). "The bleeding which frequently occurs at the time of implantation, in the monkey as well as in women, carries his name (Hartman's sign)."[31]

Hartman was also seriously committed to the development of contraception. In 1922, participants in an International Neo-Malthusian and Birth Control Conference lamented the lack of clarity about the timing of fertility,[32] and the next decade saw numerous efforts in this direction. In a long series of publications, Hartman explicitly transformed his knowledge of the menstrual cycle into a method of contraception.[33] Called the "rhythm method," it requires menstrual charting of some kind plus abstinence or the use of a barrier method during the vulnerable interlude of fertility. An array of users has long been interested, from Roman Catholics to those who cannot or will not take medications or use devices for whatever reasons. Hartman's 1933 article was titled "Catholic advice on the safe period," and was published in the *Birth Control Review*, the journal of the American Birth Control League.

The key to using this method successfully is *accurate* knowledge of the time of fertility. In the 1930s, this was deemed achievable only by timing the fertility cycle. This newly constructed female cycle was, in fact, contrary to earlier conceptualizations of the time of fertility which had often asserted that it was the time of menstruation. Thus, earlier rhythm methods of contraception had favored unprotected intercourse in the weeks just after menstruation (when it turned out pregnancy was most likely), while abstinence was advocated during menstruation when conception is least likely.

Hartman worked with Hopkins colleague Raymond Pearl on human studies. The major difficulty encountered was the range of variation of the timing of fertility both among women as a group and within individual women over time.[34] Hartman's initial summary work was *Time of Ovulation in Women: A Study on the Fertile Period in the Menstrual Cycle* (1936), one of the books in the innovative Medical Aspects of Fertility Series sponsored by the National Committee on Maternal Health, a physician-led birth control advocacy organization. His career capstone work was *Science and the Safe Period: A Compendium of Human Reproduction* in 1962. Hartman was also committed to popularizing scientific knowledge and reported that a paper he wrote, "How large is the human egg?" was rejected by *Ladies Home Journal* as too close to the pornographic! It ended up in *Scientific American*.[35] Hartman retired from the Department of Embryology in 1941, went on to the headship of the Departments of Zoology and Physiology at the University of Illinois, and served as Research Director for Ortho Pharmaceuticals.[36] One of Hartman's major contributions is the attention he paid to cyclic variation, including his comparative work.

The non-pregnant uterus and menstruation

Closely related to the female cycle work but with different emphases were the Department's several projects on the non-pregnant uterus and menstruation. Begun prior to 1925, and continuing throughout the coalescence period, this work came to be called biological or physiological (as opposed to "purely" endocrinological). Carl Hartman, George Bartelmez, his student Joseph Eldridge Markee, and others studied the menstrual cycle *per se* and related problems.

There was at this time debate that was quite heated as to the possibility of menstruation without ovulation or fertility – anovular menstruation.[37] Reportedly, "Bartelmez soon became convinced of Hartman's contention that menstruation was independent of ovulation or of pre-gravid changes of the uterine mucosa, and occurred as a normal cyclical event in which bleeding was the only constant feature. He was able to confirm Hartman's work in the rhesus monkey by demonstrating from human material that menstruation may occur in the absence of a large ovarian follicle or a corpus luteum."[38] There were several classic experiments around these problems, and considerable international competition for primacy of publication.[39] These more physiological researches led to sterility/infertility topics as a strong line of post-war research,[40] and became the basis upon which fertility control via hormonal contraception (the Pill) could later be built.

The changes occurring in the vasculature of the endometrium (uterine lining) during the menstrual cycle were also of particular interest and required mapping in both pregnant and non-pregnant females. George Bartelmez and his student Markee worked intensively on these problems both at the University of Chicago and at the CIW Department of Embryology as very regular visitors and collaborators.[41] Bartelmez first visited the Department in 1917 with a very early embryo in hand and was there for Corner's early monkey colony work, which deeply intrigued him. He and Markee became full-fledged members of "the monkey fraternity," also maintaining a small colony at Chicago funded through a Rockefeller grant to that University. Markee later moved to Stanford University.[42]

Bartelmez took his Ph.D. in Embryology from the University of Chicago in 1910, then moved over to its Department of Anatomy to develop its human embryology collection, and remained there until 1950, when he moved to the CIW Department of Embryology until *c.* 1957. In the reproductive realm, Bartelmez' focus was on the nature of the changes in the uterine mucous membrane which were responsible for menstrual bleeding. Within the broader work on the female cycle, Corner had focused on the corpus luteum and Hartman on the timing of menstruation in relation to it (discussed above). Bartelmez then correlated these luteal phases with those of the uterine lining, finding a specialized vascular supply. His initial summary statement and review framed this field.[43]

This work led Bartelmez and his student Markee to focus next on the specialized spiral arteries of the vascular system of the non-pregnant uterus in the monkey. Elizabeth Ramsey, then loosely affiliated with the Department (discussed fully below), facilitated their work with the CIW rhesus colony. Markee then did ground-breaking work that allowed the menstrual process to be visible to the human eye *in vivo*. He explanted endometrial tissue into the eyes of monkeys making the menstrual cycle visible and he observed it over several years. Markee described a rhythmic vascular constriction and relaxation of the tissue as "blushing and blanching."[44] Bartelmez theorized that hemorrhage from these vessels (menstruation) might start and stop via the "blushing and blanching" described by Markee, and demonstrated the constrictions of the spiral arteries.[45] Bartelmez then produced a second major review of menstrual cycle research.[46] "It was soon realized that the endometrial vessels of the nonpregnant cycle become the uteroplacental vessels with the onset of pregnancy."[47] Ramsey herself then joined the Department and pursued this problem of uteroplacental vascularization (discussed below). The most striking aspect here is that both Markee's and Ramsey's work yielded visualization of physiological processes never before seen live – a broad vision of anatomy indeed.

Jessie King and collaborative research on women

In another manifestation of their broad style, Mall, followed by others in the Department, had long called for cooperative efforts between laboratory and clinical researchers. In this vein, Corner's research on the menstrual cycle in rhesus monkeys stimulated study of that cycle in human females by Dr. Jessie King.[48] Dr. King had taken her Ph.D. at Cornell in 1911 and was professor of physiology at Goucher College in Baltimore, one of the rare women in the first generation of US reproductive scientists and a regular visitor/collaborator of the Department. She pursued traditional reproductive physiological and endocrinological research on animal models (especially the physiology of the uterus) as well as "clinical" research.[49] In the latter, her major focus was "normality" studies of the female cycle including periodic cardiovascular, temperature, and knee-jerk reaction variations, menstrual intervals, and their relation to vaginal (Pap) smears.[50]

Part of King's work was done in collaboration with George W. Corner, then at the University of Rochester, who had earlier studied the menstrual cycle of monkeys (noted above). Following up on Corner's monkey data, King gathered data from women describing their cycles in detail. She also got women (her colleagues at Goucher) to do daily vaginal smears.[51] While reluctant to generalize from her small series of subjects, King made several key points in her work. First, she found "surprisingly frequent variations in the length of the cycles in mature normal women." Second, she confirmed that the Pap smear "is a doubtful index of changes transpiring in the ovary and uterus," and therefore cannot be used for staging the cycle. There had been hopes that Pap smears would allow clear staging in many if not most mammals, which would have been very helpful in doing reproductive research.

Alas, studies pursued across biology, medicine, and agriculture demonstrated that this was not the case.[52] Third, King found in the literature "a tendency to overemphasize the inefficiency [difficulty in working or doing other tasks] of women during the menstrual period" as compared with her own research findings.[53] There were major debates at this historical juncture regarding menstrual debility to which King thus contributed. The debates focused on the proper status and roles of women and women's health, just as women had become entitled to vote as US citizens as the result of decades of feminist activism.

King's findings of irregularity were criticized on methodological grounds, not only for small sample size but for too old and too educated or high classed a sample. She then drew on Corner's contacts in Rochester and others' in Baltimore for access to more diverse women – from the Eastman Kodak Plant and the Western Electric Plant – spending a year in Dr. Corner's Department of Anatomy at the University of Rochester Medical School under a grant from the National Research Council.[54] Thus, her next study of menstrual intervals included subjects who were machine operators, secretaries, shop clerks, and a maid. She wrote: "It supports rather strikingly the former conclusion" about the normalcy of menstrual irregularity – differences occurring within a single subject.[55] As Hartman put it, "There are almost no regularly menstruating women, any more than there are regularly menstruating monkeys."[56]

Sperm and ovum transport

In addition to his work on the female cycle and menstruation noted above, Carl Hartman also studied sperm and ovum transport. He noted at the end of his career:

> The short viability of egg and sperm (on which I first reported 36 years ago) [Hartman 1924] . . . has been repeatedly demonstrated and it is evident that the acceptance of these facts is of great significance to gynecology, for example, for the estimation of the fertile and infertile, or so-called "safe" period of the menstrual cycle . . . I shall pass over lightly the mechanism of sperm and egg transport as I have only lately raised a number of embarrassing questions in this area [Hartman 1957]. For two centuries and up to recent years, the almost universal opinion was that sperm made their way to the site of fertilization by their own efforts. [But] A combination of tubal movements and sperm activity was clearly demonstrated . . . Within my time a compromise has been struck with regard to the relative importance of ciliary and peristaltic action. The power of the cilia in moving objects is spectacularly seen [in a film of the frog] . . . The hormonal control of the transport of the egg is illustrated strikingly by the fact that from mouse to cow, monkey and man, the ovum reaches the uterus in about 3.5 days and its arrival is correlated with the interval required for the corpus luteum to come into full activity inasmuch as progesterone hastens the migration of the egg . . .[57]

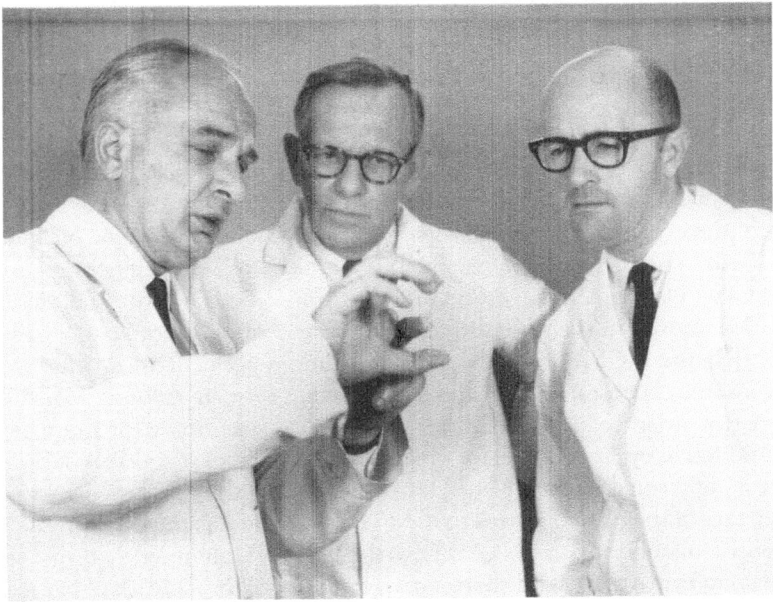

Figure 4.3 David W. Bishop (center).

George Corner had also worked on ovum transport in the sow as part of the cycle, but it was Hartman's specialty and the focus of his papers in *Sex and Internal Secretions*.[58]

In 1951, David Wakefield Bishop (Fig. 4.3), who had taken his Ph.D. at the University of Pennsylvania, joined the CIW staff and stayed until 1977. He followed in Hartman's footsteps and became a leading expert in the biology of spermatozoa, writing the major review of this area for the third edition of *Sex and Internal Secretions*.[59] He stated that, "In no way is the present review intended to represent a renovation of the comparable section in the second edition . . . (Hartman 1939). It would be both presumptuous and impracticable to attempt to update Professor Hartman's discussion of the physiologic role of spermatozoa in reproduction; this stands as a landmark now two decades old. In his review, many problems were noted, some since solved, others still in the course of solution, and many even yet ignored."[60] Hartman's tradition was thus most respectfully sustained. Bishop's particular physiological focus was on sperm motility, finding that the rhythmic movement of sperm has many of the features of muscle contraction studied by others in the Department, such as Samuel Reynolds and Arpad Csapo (discussed below). Bishop was also interested in male infertility produced by injection of testicular extract, an auto-immune response Bishop sought to identify, and which is of considerable interest today as a possible means of male contraception.[61] He retired in 1967.

No applied contraceptive research allowed

In the USA, the National Committee on Maternal Health (NCMH), a physician-sponsored birth control advocacy organization, made at least one attempt to "piggyback" applied spermicide research through the back door of "basic" sperm survival research. In 1938, the NCMH offered a grant to the CIW Department of Embryology to study the transport and viability of spermatozoa in the genital tracts of female dogs and monkeys. The plan was that the NCMH would locate and fund a scientist to do this work, cover some expenses, and use the Department's primate colony.[62] The CIW Department agreed, "provided work is designed specifically for study of the reproductive cycle and not for collateral problems of a social type . . ."[63] But in a personal letter to Carl Hartman, Raymond Squier, then Executive Secretary of the NCMH, tried to remind Hartman that another reproductive scientist member of the NCMH (Earl Engle of Columbia) had discussed this matter privately with Hartman. The NCMH thought they had come to an understanding that spermicidal testing would be incorporated into the research. Squier said he was sure that Hartman understood that the NCMH could not afford to spend "$3000 simply on further study of the estrous cycle of dogs or other work on monkeys having no relation at all to possible practical applications for the control of human reproduction."[64]

Despite his own long-term commitments to the birth control movement, Hartman's response fell fully within the strategy of reproductive scientists *vis-à-vis* their birth control audiences; he refused to piggyback the contraceptive research. He wrote to the head of the CIW Department, saying that: "I assured Squier that we could work on any phase of pure science that we wished, leaving propaganda and 'applications or social implications' for organizations like his. As to effect of chemical or physical agents on sperms – we don't propose to touch that subject unless we get a new "lead" that justifies [it] . . . What we shall do is study sperm survival under normal conditions – there will be little time for anything else."[65] Even Hartman, a former Chairman for Research of the NCMH from 1934 to 1937, would not bend the rules or cross the boundaries of the CIW Department specifically or of the basic reproductive research enterprise generally.[66]

The pregnant uterus and placenta

This line of work grew after George Corner became Director of the Department in 1940. Louis B. Flexner, nephew of the medical science pioneers and a leading anatomist, was recruited to the CIW Department by Corner around 1940 from the Hopkins Department of Anatomy. He brought modern physiological methods to the study of placental transmission and fetal physiology, emphasizing biochemistry. He also did comparative work on placental permeability, using radioactive substances as tracers. Flexner gave the Harvey Lecture in 1951, and then moved to the University of Pennsylvania as chair of the Department of Anatomy, soon becoming Director of the Neurological Institute.[67]

Another line of research on the uterus, undertaken by both Samuel R. M. Reynolds and Arpad Csapo, focused on the uterine musculature. Sam Reynolds was appointed in 1941 from Long Island Medical College. He held his Ph.D. in physiology from the University of Pennsylvania.[68] His prior work had been extensively supported by the NRC Committee for Research on Problems of Sex, with 24 publications through these grants from 1934 to 1941,[69] including a major review on uterine contractility.[70] He went on to write the definitive book *The Physiology of the Uterus* (1949), and continued work on gestational mechanisms, including endocrinological phenomena, defining multiple processes of "uterine accommodation of the products of conception."[71] His work often had the comparative edge so characteristic of the broad anatomical style of the Department.

Arpad Csapo came to the Department from Hungary via Sweden in 1949 on a fellowship arranged by Corner and in 1951 was made a permanent member of the Department. He was most knowledgeable about the chemistry and physiology of voluntary (skeletal) muscles. He then worked with George Corner on the involuntary uterine muscle and went on to a productive career at Washington University in St. Louis focused on the physiology of the uterine muscle and its role in pregnancy and labor. Corner noted: "The gist of all these researches is that progesterone reduces the transmission, throughout the uterine muscular tissue, of the impulses to contract. The clinical use of progesterone is partly based on this fundamental physiological action."[72] Csapo and his colleagues promoted the concept of an active myometrium, spontaneously contractile, which is restrained by hormone action during pregnancy. The concept remains important in terms of understanding and treating preterm labor and in the emergent specialty of fetal surgery.[73]

Elizabeth Ramsey (Fig. 4.4), a physician and pathologist, was a serious contributor in the second generation of reproductive scientists. She initially came to the CIW Department in 1934 to study the Yale embryo placed in the Department's collection. Ramsey herself had unexpectedly discovered this 14-day embryo during the autopsy of a young woman, her first autopsy. It was the earliest known human embryo at the time. She then worked – on a volunteer basis for many years – on that embryo, focusing on the

Figure 4.4   Elizabeth Ramsey.

maternal vascular system. Ramsey also became deeply involved with the primate colony during these years, greatly facilitating her own and others' research. In 1949, under Corner's directorship, she was named a research associate.[74]

By then, Ramsey's particular focus was on the development of the placenta in pregnancy and the problem of how the maternal arterial and venous bloods are kept separate in the amorphous intervillous space of the hemochorial placenta. William and John Hunter had demonstrated the independence of maternal and fetal circulations in the eighteenth century. Modern methods, including X-ray studies and her carefully staged materials, allowed her to supplant earlier theories. Here, she first worked with CIW scientist Samuel R. M. Reynolds.[75]

Later, Ramsey collaborated with George Corner Jr., the son of the previous Director, and others. They used radioangiographic techniques in a specially designed operating room.[76] She describes this work vividly:

. . . radio-opaque substances were injected into the systemic bloodstream of pregnant rhesus monkeys and their progress followed by rapid serial x-ray films and cine-radiography through the uteroplacental veins. Thus *actual visualization of placental circulation was achieved.* This work, carried out at the Carnegie Laboratory over a period of years by Ramsey and her group in collaboration with M. W. Donner of the Johns Hopkins Radiology Department, was based on the Carnegie monkey colony. The work has been confirmed elsewhere. The physiological concept states that the . . . maternal systemic blood pressure . . . drives the incoming blood in fountain-like spurts toward the chorionic plate, thus preventing shortcutting to adjacent venous exits . . .[77]

This was the culmination of Ramsey's research. Her now classic model of placental blood flow was published in 1961 (Fig. 4.5). Dr. Ramsey retired in 1971, but she remained a sought-after international lecturer whose capstone books remain significant.[78] Her clinical contributions include establishing the concept of the "maternal–placental–fetal complex," paving the way for understanding "abruptio placentae," a leading cause of perinatal death, and other problems of pregnancy. A large literature has emerged on the separation of the fetus and mother known as "maternal/fetal conflict." Ramsey's, as well as Corner and Bartelmez's, research can be viewed as early articulations of that separation.[79]

Working the flipside of Ramsey's project, staff member Bent Boving sought to understand the process of implantation, which requires a receptive uterine wall, an adhesive interaction between the wall and the blastocyst, and finally successful invasion of the trophoblast into the uterine wall. The penetration by the trophoblast is in the immediate vicinity of a maternal subepithelial capillary ultimately to allow placental development and blood flow. Sustaining the Department's visualization tradition, Boving also developed methods that allowed observations inside the living uterus.[80] He moved to the Department of Anatomy at Wayne State University Medical School in 1968.

Primate fetal sexual development

Robert Kyle Burns was a student of renowned embryologist Ross Harrison at Yale who had taught at the University of Cincinnati. His research in experimental embryology centered on the development and differentiation of the reproductive organs, particularly the testis, ovary, and their ducts. He had worked with Corner at Rochester and came with him to the Department of Embryology in 1940. He borrowed a page from Carl Hartman's book and switched from using amphibian larvae to studying the embryology of the opossum; his work was greatly facilitated by its being a pouched mammal.[81] He successfully induced sex reversal in the gonads and genital tract of

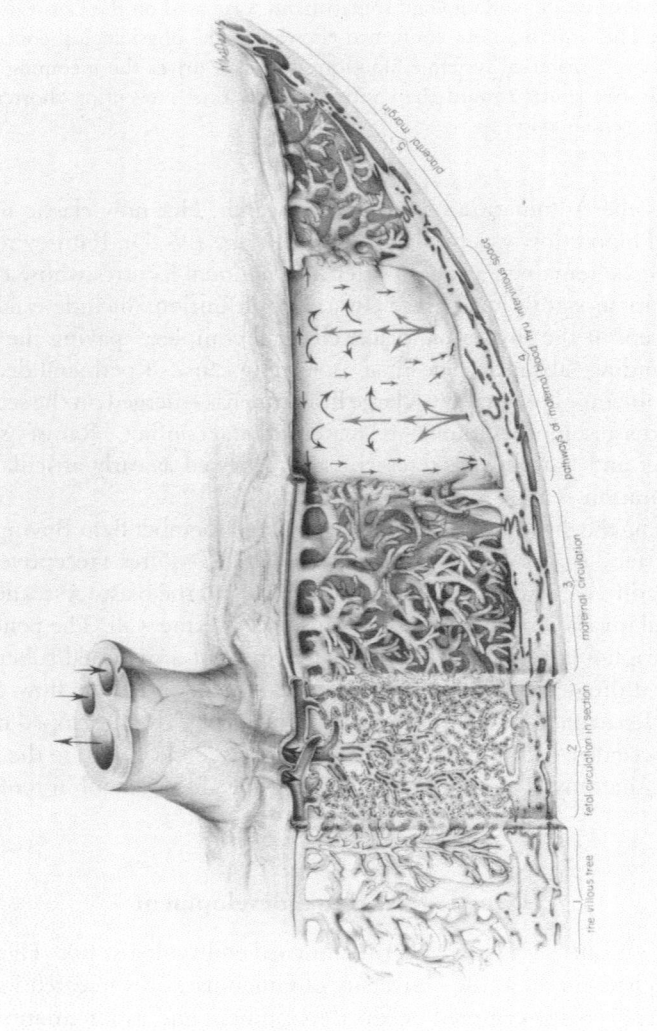

Figure 4.5   Elizabeth Ramsey's placental circulation diagram.

opossum embryos experimentally through the use of hormones.[82] Robert Burns was a distinguished reproductive biologist and retired from the Department in 1962.

Primate fetal pathology studies

Some reproductive research had broad cultural symbolic import as well as medical consequences. Research in fetal pathology was one such area. Corner, with his usual modesty, describes research on abnormal monkey embryos:

> Bartelmez and I joined in a little monograph [1954] describing nine very early abnormal embryos of the rhesus monkey from the Carnegie colony . . . [I]t is unique in the literature of primate embryology in that we could study, along with the embryos, the maternal ovaries and endometrium (lining of the uterus). The findings helped to lay to rest the old idea that embryonic pathology is always caused by uterine inflammation [a problem in the mother]. In five of our cases the ovaries and endometrium were normal; the embryonic abnormality must have resulted from constitutional defects of the embryo itself. In other cases, the possibility existed that the abnormality of the embryo resulted from primary failure of the corpus luteum.[83]

Here, a reproductive scientist made a concrete research offering to obstetricians, gynecologists, and animal agriculturalists. That there could be serious pathology of the embryo/fetus in a healthy mother was a radical departure from the canon. It had been assumed to be impossible. The research thus symbolically "absolved" women from "blame" for the first time for at least some proportion of fetal abnormalities and fetal loss.

Conclusions

The current Director of the Department of Embryology, Alan Spradling, has described departmental philosophy as follows: "Scientific leadership requires exceptional individuals with the insight, resources, and courage to investigate the margins of what is feasible and respectable . . . Bold ventures are encouraged . . ."[84] His statement demonstrates the continuity of daring encouraged there. Certainly, in the early decades of the Department, the reproductive sciences were marginal at best. Some might find them even more controversial today.[85] Clearly, the Department of Embryology located at Johns Hopkins Medical School was profoundly daring in its support of this line of scientific work for so many decades.

In this brief chapter, I have offered an overview of the research on reproduction undertaken at the Department. I have mapped the key areas of research but, due to space limitations, have not elaborated deeply on the science *per se* and how it fits within the discipline more broadly. Several major reviews of the field undertaken in the 1960s and 1970s provide such detailed elaborations, and I refer readers to them.[86]

Table 4.1   *Carnegie Contributions to Embryology*[87]

Articles on reproductive phenomena 1915–1966	
Placenta	16
Ova	9
Ovary	9
Menstrual cycle	4
Uterus	4
Phallus	3
Urogenital sinus	2
Endometrium	2
Fallopian tube	1
Total	50

During the decades covered here, 1913–71, the main goal of reproductive researchers *vis-à-vis* their discipline and professions was to put reproductive research "on the map" as a fully scientific, appropriately experimental, appropriately physiological and later biochemical/endocrinological endeavor. Reigning paradigms and emergent standards of scientific research had to be applied in full, systematically and routinely, in reproductive research to earn it stature as a fully scientific field of endeavor. Coalescence of the field thus also included the usual activities of professionalization of a new discipline, publishing new journals, forming new associations, and having national and international meetings. The CIW Department of Embryology was a major contributor, not only sponsoring the broad array of research by its staff and collaborators discussed above, but also through sponsoring its journal, *Carnegie Contributions to Embryology*, published from 1915 to 1966. This journal was a major venue for publishing reproductive as well as embryological research, including fifty such papers by authors included here and others (Table 4.1).

The applications of the reproductive science pursued in the Department and beyond were varied and many innovations were built upon this work in the Department. The benefits of the reproductive sciences to medical scientists and to clinical medicine were multi-fold, focused in two distinctive directions. First, it offered a highly scientific line of work to medicine as it sought to establish scientific medicine as the reigning medical paradigm. Second, it offered an extensive array of information and therapeutics to obstetricians, gynecologists, and urologists (who addressed male reproductive phenomena) for use in both the diagnosis and treatment of functional (physiological) reproductive problems. These were among the very first real alternatives to surgery available as therapeutics in these specialties.

In terms of diagnostic and therapeutic offerings to about 1940, Leonardo offered a list of twenty-one items including: an early pregnancy test, a test for hydatid moles (a serious complication of pregnancy), a hormone treatment for gonorrheal vulvovaginitis in children, a method of determining whether ovulation has occurred (endometrial biopsy), an understanding of anovular menstruation, potential hormonal treatments for lactation, prevention of miscarriage, absence of menstruation and postpartum hemorrhage, treatments for dismenorrhea, and an understanding of mittelschmertz (intermenstrual pain) as due to ovulation.[87] Corner also clarified that miscarriages are commonly due to problems of the fetus rather than maternal pathology. While Jessie King established that the vaginal smear could not be used to stage the human female cycle, Papanicolaou did extensive work toward using it in the diagnosis of cervical and uterine malignancy. Another Mall student and friend of the Department, Herbert M. Evans, clarified the role of vitamins in reproductive processes including pregnancy. Hartman clarified the timing of ovulation and fertility in women and published this as "Catholic advice on the safe period." Diagnostics and therapeutics for fertility and sterility became important medical offerings.[88] Last, Elizabeth Ramsey's research indicated that unduly prolonged contractions in labor place the fetus in considerable jeopardy due to curtailed placental circulation.

A significant proportion of this productive research on reproductive phenomena was carried out by the staff of the CIW Department of Embryology and their impressive collaborators over the period of the heyday of the reproductive sciences there, 1913–71. The infrastructural investment in one of the first monkey colonies for biomedical research by the Carnegie Foundation clearly paid off. Pioneering work in visualization of physiological processes was accomplished. A significant number of the CIW reproductive scientists were elected to membership in the National Academy of Sciences (Bartelmez, Burns, Corner, Flexner, Hartman, Mall, and Schultz), both honoring the Department and legitimizing their work.[89] The Department of Embryology was indeed one of the major centers of medically oriented reproductive science of the twentieth century.

Acknowledgements

Much of the data presented here draws on a larger project on the history of the US reproductive sciences.[90] I am grateful to the Rockefeller University for a grant to visit the Rockefeller Archives, and to various grants from the University of California, San Francisco, that permitted archival research at the Johns Hopkins University. Special thanks to Siobhan Harlow for graciously hosting me during that archival research so many years ago. I am especially indebted also to Toby Appel, who very generously loaned me her complete files on Jessie L. King, and to Baleen Shemirani for research assistance.

Notes

1. See CIW Archives: "Resume of Taped Interview with Dr. Elizabeth Ramsey, 11 January 1978," pp. 7–8.
2. Adele E. Clarke, "Embryology and the development of American reproductive sciences, 1910–1945," in Ronald Rainger, Keith Benson, and Jane Maienschein (eds.), *The American Expansion of Biology* (New Brunswick, NJ: Rutgers University Press, 1991), pp. 107–32; Adele E. Clarke, *Disciplining Reproduction: Modernity, American Life Sciences and the "Problems of Sex"* (Berkeley, CA: University of California Press, 1998), especially ch. 3; and Adele E. Clarke, "Maverick reproductive scientists and the production of contraceptives *c.* 1915–2000," in Anne Saetnan, Nelly Oudshoorn, and Marta Kirejczyk (eds.), *Bodies of Technology: Women's Involvement with Reproductive Medicine* (Columbus, OH: Ohio State University Press, 2000), pp. 37–89.
3. See, e.g., James D. Ebert, "Department of Embryology report – introduction," Carnegie Institution of Washington *Year Book* 75 (Washington, DC: Carnegie Institution of Washington, 1975–6).
4. Adele E. Clarke, "Research materials and reproductive science in the United States, 1910–1940," in Gerald L. Geison (ed.), *Physiology in the American Context, 1850–1940* (Bethesda, MD: American Physiological Society, 1987), pp. 323–50. Reprinted in S. Leigh Star (ed.), *Ecologies of Knowledge: New Directions in Sociology of Science and Technology* (Albany, NY: SUNY Press, 1995), pp. 183–219.
5. Diana Long Hall, "Sex, fertility and taboo: The committee for research on problems of sex, 1920–1940," unpublished manuscript, provided by the author; presented at Workshop on Historical Perspectives on the Scientific Study of Fertility in the United States, American Academy of Arts and Sciences, 1978; Adele E. Clarke, "Controversy and the development of American reproductive sciences," *Social Problems* 37(1) (1990), pp. 18–37. Reprinted in Andrea Tone (ed.), *Controlling Reproduction: An American History* (Wilmington, DE: Scholarly Resources Inc., 1997); and Clarke, *Disciplining*, especially ch. 8.
6. For a case study of a biological center, see Adele E. Clarke, "Money, sex and legitimacy at Chicago, 1900–1940: Lillie's center of reproductive biology," Special Issue on Biology at the University of Chicago, *Perspectives on Science* 1(3) (1993), pp. 367–415.
7. Marrian noted the "heroic age"; see A. S. Parkes, "The rise of Reproductive physiology, 1926–1940: The Dale lecture for 1965," *Endocrinology* (*Proceedings of the Society*) (1966), p. xx. Parkes called it the "endocrinological gold rush." See A. S. Parkes, "Prospect and retrospect in the physiology of reproduction," *British Medical Journal* (14 July 1962), p. 72; Edgar Allen (ed.), *Sex and Internal Secretions*, 1st edn. (Baltimore, MD: Williams and Wilkins, 1932); Edgar Allen (ed.), *Sex and Internal Secretions*, 2nd edn. (Baltimore, MD: Williams and Wilkins, 1939); and William C. Young (ed.), *Sex and Internal Secretions*, 3rd edn. (Baltimore, MD: Williams and Wilkins, 1961); George W. Corner, "Foreword," in William C. Young (ed.), *Sex and Internal Secretions*, 3rd edn. (Baltimore, MD: Williams and Wilkins, 1961), pp. ix–xii.
8. Compare Clarke, *Disciplining*, ch. 7, with Roy O. Greep, M. A. Koblinsky, and F. S. Jaffe, *Reproduction and Human Welfare: A Challenge to Research* (Boston, MA: MIT for the Ford Foundation, 1976), pp. 367–71.

9. Florence R. Sabin, *Franklin Paine Mall* (Baltimore, MD: Johns Hopkins University Press, 1934); Lawrence D. Longo, "Obstetrics and gynecology," in Ronald L. Numbers (ed.), *The Education of American Physicians* (Berkeley, CA: University of California Press, 1980), pp. 218–19; Lawrence D. Longo, "John Whittridge Williams and academic obstetrics in America," *Transactions and Studies of the College of Physicians of Philadelphia* Series V, III (4) (1981), pp. 221–54; and R. Hahn, "Division of labor: obstetricians, women, and society in Williams Obstetrics, 1903–1985," *Medical Anthropology Quarterly* (NS) 1(3) (1987), pp. 256–82.

10. James Reed, *The Birth Control Movement and American Society: From Private Vice to Public Virtue*, 2nd edn. (Princeton, NJ: Princeton University Press, 1983), pp. 163–4.

11. Longo, "Obstetrics and gynecology," p. 223.

12. I. D. Raacke, "Herbert McLean Evans (1882–1971): a biographical sketch," *Journal of Nutrition* 113(4) (1983), p. 931.

13. George W. Corner, *Anatomist at Large: An Autobiography and Selected Essays* (New York: Basic Books, 1958), pp. 27, 30; George W. Corner, "Autobiographical sketch for the files of the National Academy of Science," CIW Archives, unpublished manuscript, December 1953.

14. John G. Gruhn and Ralph R. Kazer, *Hormonal Regulation of the Menstrual Cycle: The Evolution of Concepts* (New York: Plenum Medical Book Company, 1989).

15. Corner, *Anatomist at Large*, p. 30. See also, George W. Corner, *The Seven Ages of a Medical Scientist: An Autobiography* (Philadelphia, PA: University of Pennsylvania Press, 1981); Corner, "Autobiographical Sketch"; and Carl G. Hartman, "The Scientific achievements of George Washington Corner, M.D.," *American Journal of Anatomy* 98(1) (1956), pp. 5–19.

16. Robert E. Kohler, *Partners in Science: Foundations and Natural Scientists, 1900–1945* (Chicago, IL: The University of Chicago Press, 1991).

17. Sabin, *Mall*, p. 303.

18. Ebert, "Department of Embryology," p. 7.

19. Corner, *Seven Ages*, p. 287.

20. Rosemary Stevens, *American Medicine and the Public Interest* (New Haven, CT: Yale University Press, 1971), p. 162.

21. The American Foundation (ed.), *Medical Research: A Midcentury Survey*. Vol. II. *Unsolved Clinical Problems in Biological Perspective* (Boston, MA: Little Brown for the American Foundation, 1955), pp. 139–40; and Diana E. Long, "Physiological identity of American sex researchers between the two world wars," in Gerald L. Geison (ed.), *Physiology in the American Context, 1850–1940* (Bethesda, MD: American Physiological Society, 1987), pp. 263–78.

22. Jane Maienschein, "Epistemic styles in German and American embryology," *Science in Context* 4(2) (1991), pp. 407–27; Jane Maienschein, *Transforming Traditions in American Biology, 1880–1915* (Baltimore, MD: Johns Hopkins University Press, 1991); Gregg Mitman, Adele Clarke, and Jane Maienschein (guest eds.), "Introduction to Special Issue on Biology at the University of Chicago, c. 1891–1950," *Perspectives on Science* 1(3) (1993), pp. 359–66; Keith R. Benson, "American morphology in the late nineteenth century: the Biology Department at Johns Hopkins University," *Journal of the History of Biology* 18(2) (1985), pp. 163–205; and Scott F. Gilbert, "The embryological origins of the gene theory," *Journal of the History of Biology* 11 (1978), p. 308.

23. Philip J. Pauly, "The appearance of academic biology in late nineteenth century America," *Journal of the History of Biology* 17(3) (1984), p. 381.

24. Sabin, *Mall*, pp. 255–8. See also John B. Blake, "Anatomy," in Ronald L. Numbers (ed.), *The Education of American Physicians* (Berkeley, CA: University of California Press, 1980), p. 41; and George W. Corner, "The past of anatomy in the United States," *Anatomical Record* 137 (1960), p. 181.

25. Raacke, "Evans," p. 931, emphasis added.

26. T. Dale Stewart, "Adolph Hans Schultz," *Biographical Memoirs of the National Academy of Sciences, USA* 54 (1983), pp. 329–31.

27. Hartman, quoted in Rudolph P. Vollman (ed.), *Fifty Years of Research on Mammalian Reproduction: A Bibliography of the Scientific Publications of Carl G. Hartman* (Washington, DC: USDHEW, Public Health Service Publication No. 1281, 1965), p. vii.

28. John D. Biggers, "Introduction of the First Carl G. Hartman Lecturer," *Biology of Reproduction* 2 (1970), pp. 1–4; and Vollman "Carl Hartman's contributions to the Physiology of Reproduction," pp. v–vii in Rudolph P. Vollman (ed.), *Fifty Years of Research on Mammalian Reproduction: A Bibliography of the Scientific Publications of Carl G. Hartman* (Washington, DC: USDHEW, Public Health Service Publication no. 1281, 1965).

29. Corner, *Seven Ages*, p. 164; for the research, see George W. Corner, "Ovulation and menstruation in *Macacus rhesus*," *Carnegie Contributions to Embryology* 75 (1923), pp. 73–110; George W. Corner and Willard M. Allen, "Physiology of the corpus luteum 2. Production of a special uterine reaction (progestational proliferation) by extracts of the corpus luteum," *American Journal of Physiology* 88 (1929), pp. 326–9; George W. Corner and Willard M. Allen, "Physiology of the corpus luteum 3. Normal growth and implantation of embryos after very early ablation of the ovaries, under the influence of extracts of the corpus luteum," *American Journal of Physiology* 88 (1929), pp. 340–6; and George W. Corner, "The nature of the menstrual cycle," *Medicine* 12 (1933), pp. 61–82.

30. Carl G. Hartman, "Bimanual rectal palpation as applied to the female rhesus monkey," *Anatomical Record* 45 (1930), p. 263.

31. Vollman, "Carl Hartman," p. vii; see also Carl G. Hartman, "Observations on the viability of the mammalian ovum," *American Journal of Obstetrics and Gynecology* 7 (1924), pp. 40–3; Carl G. Hartman, "On the relative sterility of the adolescent organism," *Science* 74 (1931), pp. 226–7; Carl G. Hartman, "Studies in the reproduction of the monkey, *Macacus (Pithecus) rhesus* with special reference to menstruation and pregnancy," *Carnegie Contributions to Embryology* 134 (Washington, DC: Carnegie Institution of Washington, 1932), pp. 1–161; Carl G. Hartman, "Ovulation and the transport and viability of ova and sperm in the female genital tract," in Edgar Allen (ed.), *Sex and Internal Secretions* (Baltimore, MD: Williams and Wilkins, 1932), pp. 647–732; Carl G. Hartman, "Catholic advice on the safe period," *Birth Control Review* 17 (May, 1933), pp. 117–19; Carl G. Hartman, *Time of Ovulation in Women* (Baltimore, MD: Williams and Wilkins, 1936); Carl G. Hartman, "Facts and fallacies of the safe period," *Journal of Contraception* 2 (1937), pp. 51–61; and Carl G. Hartman, "Studies on reproduction in the monkey and their bearing in gynecology and anthropology," *Endocrinology* 25 (1939), pp. 670–82.

32. Raymond Pierpoint, *Report of the Fifth International Neo-Malthusian and Birth Control Conference* (London: William Heinemann, 1922), p. 270.

33. Hartman, "Catholic Advice;" Hartman, *Time of Ovulation*; Hartman, "Facts;" and Carl G. Hartman, *Science and the Safe Period: A Compendium of Human Reproduction* (Baltimore, MD: Williams and Wilkins, 1962).

34. Hartman, "Studies;" Hartman, *Science*, p. vii; and Raymond Pearl, "Contraception and fertility in 2000 women," *Human Biology* 4 (1932), pp. 363–407. See also Garland E. Allen, "Old wine in new bottles: from eugenics to population control in the work of Raymond Pearl," in Ronald Rainger, Keith Benson, and Jane Maienschein (eds.), *The American Expansion of Biology* (New Brunswick, NJ: Rutgers University Press, 1991), pp. 231–61.

35. Carl G. Hartman, "Research should spell FUN," in G. W. Duncan, R. J. Ericsson, and R. G. Zimbelman (eds.), *Capacitation of Spermatozoa and Endocrine Control of Spermatogenesis* (Oxford, UK: Blackwell Scientific Publications, 1967), p. 6.

36. Vollman, "Carl Hartman," p. v–xii. On the "Hartman Research Library" at Ortho, see Carl G. Hartman, "Annotated list of published reports on clinical trials with contraceptives," *Fertility and Sterility* 10(2) (1959), pp. 177–89.

37. See, e.g., Corner, "The nature," pp. 61–82; Corner, *Seven Ages*, p. 165–7; Hartman, "Relative sterility," pp. 226–7; and Hartman, "Studies," pp. 670–82.

38. David Bodian, "George William Bartelmez, 1885–1967," *Biographical Memoirs of the National Academy of Science, USA* XLIII (New York: Columbia University Press, 1973), p. 12.

39. For the classic experiments, see Corner, "Ovulation," and Hartman, "Studies," and "Ovulation." On the competition for primacy, see Hartman, *Science and the Safe Period*, pp. 235–52; and Vollman, "Carl Hartman." See also Paula Viterbo, *Counting the Days: A History of Natural Birth Control in America* (Cambridge, MA: Harvard University Press, forthcoming).

40. American Foundation (ed.), *Medical Research*, pp. 135–98.

41. George W. Bartelmez with C. Cuthbertson, "Histologic studies of menstruating mucus membrane of the human uterus," *Carnegie Contributions to Embryology* 24 (No. 142) (1933), pp. 141–86; and George W. Bartelmez, "Menstruation," *Physiological Reviews* 17(1) (1937), pp. 28–72.

42. Bodian, "Bartelmez," pp. 10–14. See J. Eldridge Markee, "Menstruation in intraocular endometrial transplants in the rhesus monkey," *Carnegie Contributions to Embryology* 177 (1940), pp. 220–308.

43. Bartelmez, "Menstruation."

44. Markee, "Menstruation," p. 223. Markee had numerous grants from the National Research Council's Committee for Research on Problems of Sex.

45. George W. Bartelmez, "Premenstrual and menstrual ischemia and the myth of endometrial arteriovenous anastomoses," *American Journal of Anatomy* 98 (1956), pp. 69–95. See also George W. Bartelmez, George W. Corner, and Carl G. Hartman, "Cyclic changes in the endometrium of the rhesus monkey," *Carnegie Contributions to Embryology* 34 (1951), pp. 99–144.

46. George W. Bartelmez, "The phases of the menstrual cycle and their interpretation in terms of the pregnancy cycle," *American Journal of Obstetrics and Gynecology* 74 (1957), pp. 931–1055. See also George W. Corner, "Our knowledge of the menstrual cycle, 1910–1950," Fourth Annual Addison Lecture, Delivered at Guy's Hospital, London, 13 July 1950; *The Lancet* (28 April 1951), pp. 919–23.

47. Ramsey writing in Ronan O'Rahilly, "One hundred years of human embryology," *Issues and Reviews in Teratology* 4 (1988), p. 109.

48. Corner "Ovulation," and Jessie L. King, "Menstrual records and vaginal smears in a selected group of normal women," *Carnegie Contributions to Embryology* 95 (1926), pp. 79–93.

49. King, "Menstrual records." See Jacques Cattell (ed.), *American Men of Science II: Biological Sciences* (Lancaster, PA: The Science Press, and New York: R. R. Bowker Company, 1955), p. 613.

50. On normality studies, see Suzanne W. Fletcher, Robert H. Fletcher, and M. Andrew Greganti, "Clinical research trends in general medical journals, 1946–1976," in Edward B. Roberts, Robert I. Levy, Stan N. Finkelstein, Jay Moscowitz, and Edward J. Sondik (eds.), *Biomedical Innovation* (Cambridge, MA: MIT Press, 1981), pp. 284–300. See Jessie L. King, "Concerning the periodic cardiovascular and temperature variations in women," *American Journal of Physiology* 34 (1914), pp. 202–19; Jessie L. King, "Possible periodic variations in the extent of the knee-jerk in women," *American Journal of Physiology* 42 (1916/17), p. 607; Jessie L. King, "Possible periodic variations in the extent of the knee-jerk in women," *American Journal of Physiology* 47 (1918/19), pp. 404–9; King, "Menstrual records;" and Jessie L. King, "Menstrual intervals," *American Journal of Obstetrics and Gynecology* 25 (4) (1933), pp. 583–90. For critique of methods used in cycle research on humans, see Leslie B. Arey, "The degree of normal menstrual irregularity," *American Journal of Obstetrics and Gynecology* 37 (1939), pp. 12–29.

51. King, "Menstrual records," p. 88.

52. King, "Menstrual records," p. 92. See Clarke, *Disciplining*, pp. 82–5.

53. King, "Menstrual records," p. 93. See, e.g., Sioban Harlow, "Function and dysfunction: an historical critique of the literature on menstruation and work," *Health Care for Women International* 7 (1986), pp. 39–50; and Adele E. Clarke, "Women's health over the life cycle" in Rima Apple (ed.), *The History of Women, Health and Medicine in America: An Encyclopedic Handbook* (New York: Garland Press, 1990), pp. 3–39.

54. Bessie L. Moses, "Dr. Jessie L. King," *Goucher Alumnae Quarterly* XXV (4, August 1947), p. 17.

55. King, "Concerning," p. 218; and "Menstrual intervals," p. 583.

56. Hartman quoted in Margaret Sanger (ed.), *Biological and Medical Aspects of Contraception, American Conference on Birth Control and National Recovery* (Washington, DC: National Committee on Federal Legislation for Birth Control, Inc., 1934), p. 53.

57. Carl G. Hartman, "The 1960 Ayerst Lecture: a half century of research in reproductive physiology," *Fertility and Sterility* 12 (1961), pp. 1–19 (reprinted in Vollman, *Fifty Years*, pp. 11–12); see also Carl G. Hartman, "How do sperms get into the uterus?," *Fertility and Sterility* 8 (1957), pp. 403–27; Hartman, "Catholic advice;" and Hartman, "Facts and fallacies."

58. Corner, *Seven Ages,* p. 160–1; Hartman, "Ovulation;" and Hartman, "Studies."

59. Corner, *Seven Ages,* p. 287; and David W. Bishop, "Biology of spermatozoa," in William C. Young (ed.), *Sex and Internal Secretions,* 3rd edn. (Baltimore, MD: Williams and Wilkins, 1961), pp. 707–96.

60. Bishop, "Biology," pp. 707–8.

61. Clarke, "Maverick reproductive scientists."

62. Johns Hopkins Archives (JHA) Carnegie Department of Embryology (CDoE) Streeter Papers (SP): To Dr. Merriam from G. Streeter, 12 March 1938. JHA CDoE SP (then unsorted; now likely filed under NCMH).

63. To George Streeter, Director of the Department of Embryology from President Merriam of the Carnegie Institution, 4 June 1938. JHA CDoE SP (then unsorted; now likely filed under NCMH).

64. To Hartman from Dr. Raymond Squier, 7 July 1938. JHA CDoE SP (then unsorted; now likely filed under NCMH).

65. Hartman to George Streeter, 13 July 1938. JHA CDoE SP (then unsorted; now likely filed under NCMH).

66. Rockefeller Archives Center (RAC) Rockefeller Foundation (RF) Papers: To Alan Gregg, RF, from Dr. Louise F. Bryant, NCMH, 21 December 1934. Letter from Hartman to Bryant, as attachment, 5 December 1934. RAC RF RG1.1 S200A B99 F1199.

67. Cattell (ed.), *American Men*, p. 356; and James M. Sprague, "Louis Barkhouse Flexner, January 7, 1902–March 29, 1996," *Biographical Memoirs of the National Academy of Sciences, USA* (1998), pp. 3–17.

68. Corner, *Seven Ages*, p. 285. See Cattell (ed.), *American Men*, p. 929.

69. Clarke, *Disciplining*, p. 283.

70. Samuel R. M. Reynolds, "The nature of uterine contractility: a survey of recent trends," *Physiological Review* 17(2) (1937), pp. 304–34.

71. Samuel R. M. Reynolds, *The Physiology of the Uterus* (New York: Paul B. Hoeber Pub., 1949); Samuel R. M. Reynolds, "Gestation mechanisms," in Otto V. St. Whitelock (ed.), *Special Issue: The Uterus, Annals of the New York Academy of Sciences* 75, Art. 2 (1959), pp. 691–9; and Samuel R. M. Reynolds, "The umbilical cord," in Caryl P. Haskins (ed.), *The Search for Understanding: Selected Writings of the Carnegie Institution* (Cambridge, MA: Massachusetts Institute of Technology Press, 1967), pp. 181–90.

72. Corner, *Seven Ages*, pp. 245, 293–4. See A. Csapo, "Defence mechanism of pregnancy," in G. E. W. Wolstenholme and Margaret P. Cameron (eds.), *Progesterone and the Defence Mechanism of Preganancy*, Ciba Study Group No. 9 in honor of George W. Corner (Boston, MA: Little Brown, 1961), pp. 3–26; A. I. Csapo, "The 'seesaw' theory of the regulatory mechanism of pregnancy," *American Journal of Obstetrics and Gynecology* 121(4) (1975), pp. 578–81.

73. D. G. Porter and C. A. Finn, "The biology of the uterus," in Roy O. Greep and Marjorie A. Koblinsky (eds.), *Frontiers in Reproduction and Fertility Control: A Review of the Reproductive Sciences and Contraceptive Development* (Boston, MA: MIT for the Ford Foundation, 1977), pp. 146–7; on fetal surgery, see Monica Casper, *The Making of the Unborn Patient: A Social Anatomy of Fetal Surgery* (New Brunswick, NJ: Rutgers University Press, 1998).

74. Corner, *Seven Ages*, p. 286; L. D. Longo and G. Meschia, "Elizabeth M. Ramsey and the evolution of ideas of uteroplacental blood flow and placental gas exchange," *European Journal of Obstetrics, Gynecology and Reproductive Biology* 90(2) (2000), pp. 129–33; Elizabeth M. Ramsey, "The Lockyer embryo: an early human embryo in situ," *Carnegie Contributions to Embryology* 26 (1937), pp. 99–120; and Elizabeth M. Ramsey, "The Yale embryo," *Carnegie Contributions to Embryology* 27 (1938), pp. 67–84.

75. E. C. Gillespie, E. M. Ramsey, and S. R. M. Reynolds, "The pattern of uterine growth during pregnancy in monkeys as shown in an x-ray study," *American Journal of Obstetrics and Gynecology* 58 (1949), pp. 758–64.

76. Elizabeth M. Ramsey, G. W. Corner, Jr., M. W. Donner, and H. Stran, "Radio-angiographic studies of circulation in the maternal placenta of the rhesus monkey:

preliminary report," *Proceedings of the National Academy of Science USA* 46 (1960), pp. 1003–8; Elizabeth M. Ramsey, G. W. Corner, Jr., and M. W. Donner, "Cineradiographic visualization of venous drainage of the primate placenta in vivo," *Science* 141 (1963), pp. 909–10; and Elizabeth M. Ramsey, G. W. Corner, Jr., and M. W. Donner, "Serial and cineradiographic visualization of maternal circulation in the primate (hemochrial) placenta," *American Journal of Obstetrics and Gynecology* 86 (1963), pp. 213–25.

77. Ramsey, writing in O'Rahilly, "One hundred years," p. 110, emphasis added. See also Elizabeth M. Ramsey, "The Carnegie monkey colony and the placental circulation of primates,"Carnegie Institution of Washington *Year Book* 70 (1970–1), pp. 84–93.

78. Elizabeth M. Ramsey, *The Placenta: Human and Animal* (New York: Praeger, 1982); and Elizabeth M. Ramsey and M. W. Donner, *Placental Vasculature and Circulation* (Stuttgart: Georg Thieme, 1980).

79. Longo and Meschia, "Elizabeth M. Ramsey"; T. K. A. B. Eskes, "Abruptio placentae – a "classic" dedicated to Elizabeth Ramsey," *European Journal of Obstetrics, Gynecology, and Reproductive Biology* 75(1) (1997), pp. 63–70; and in CIW Archives: "Elizabeth Maplesden Ramsey, 1906–1993"; "Resume of taped interview with Dr. Elizabeth Ramsey, January 11, 1978"; and "Dr. Elizabeth Ramsey: a profile." See also Barbara Duden, *Disembodying Women: Perspectives on Pregnancy and the Unborn*, translated by Lee Hoinacki (Cambridge, MA: Harvard University Press, 1993); Rosalind Pollack Petchesky, "Fetal images: the power of visual culture in the politics of reproduction," *Feminist Studies* 13 (1987), pp. 263–92; and Casper, *The Making*.

80. Bent Boving, "Biomechanics of Implantation," in R. J. Blandau (ed.), *The Biology of the Blastocyst* (Chicago, IL: University of Chicago Press, 1971), pp. 423–42; and Bent Boving, "Observing inside the living uterus," in J. C. Daniel, Jr. (ed.), *Methods in Mammalian Embryology* (San Francisco, CA: W. H. Freeman, Co., 1971).

81. Corner, *Seven Ages,* pp. 214, 289; see also on the opossum as research material, Clarke, "Research materials."

82. Robert Kyle Burns, "The role of hormones in the differentiation of sex," in William C. Young (ed.), *Sex and Internal Secretions*, 3rd edn. (Baltimore, MD: Williams and Wilkins, 1961), pp. 76–160.

83. Corner, *Seven Ages,* pp. 353–4.

84. Allan C. Spradling, "The Carnegie Institution of Washington, Department of Embryology," *Molecular Medicine* 3(7) (1997), p. 417.

85. Clarke, *Disciplining,* ch. 8.

86. Young, *Sex and Internal Secretions*; Greep *et al.*, *Reproduction*; and Roy O. Greep and Marjorie A. Koblinsky (eds.), *Frontiers in Reproduction and Fertility Control: A Review of the Reproductive Sciences and Contraceptive Development* (Boston, MA: MIT for the Ford Foundation, 1977).

87. This table was created by analyzing the "Index of authors, Volumes I–XXXVIII, 1915–1966," *Carnegie Contributions to Embryology,* pp. 125–31. Special thanks to Baleen Shemirani for assistance.

88. Richard A. Leonardo, *History of Gynecology* (New York: Froben, 1944), pp. 374–6.

89. E. C. Amoroso and G. W. Corner, "Herbert McLean Evans, 1882–1971," *Biographical Memoirs of Fellows of the Royal Society* 21 (1975), pp. 83–186; Edward Reynolds and Donald Macomber, *Fertility and Sterility in Human Marriages* (Philadelphia, PA: W.B. Saunders, 1924); and American Foundation (ed.), *Medical Research*.

90. See listings of NAS members on the NAS website: www.nas.edu. See also George
W. Corner, "The generality and the particularity of man," in Caryl P. Haskins
(ed.), *The Search for Understanding: Selected Writings of the Carnegie Institution*
(Cambridge, MA: Massachusetts Institute of Technology Press, 1967), pp. 211–34.
91. Clarke, "Research materials"; "Controversy"; "Embryology"; "Money"; *Disciplining*; and "Maverick Reproductive Scientists."

Bibliography

Allen, Edgar (ed.), *Sex and Internal Secretions*, 1st edn. (Baltimore, MD: Williams and
Wilkins, 1932).

(ed.), *Sex and Internal Secretions*, 2nd edn. (Baltimore, MD: Williams and Wilkins,
1939).

Allen, Garland E., "Old wine in new bottles: from eugenics to population control
in the work of Raymond Pearl," in Ronald Rainger, Keith Benson, and Jane
Maienschein (eds.), *The American Expansion of Biology* (New Brunswick, NJ:
Rutgers University Press, 1991), pp. 231–61.

American Foundation, The (ed.), *Medical Research: A Midcentury Survey*. Vol. II.
Unsolved Clinical Problems in Biological Perspective. (Boston, MA: Little Brown
for the American Foundation, 1955).

Amoroso, E. C. and G. W. Corner, "Herbert McLean Evans, 1882–1971," *Biographical
Memoirs of Fellows of the Royal Society* 21 (1975), pp. 83–186.

Arey, Leslie B., "The degree of normal menstrual irregularity," *American Journal of
Obstetrics and Gynecology* 37 (1939), pp. 12–29.

Bartelmez, George W., "Menstruation," *Physiological Reviews* 17(1) (1937), pp. 28–72.

"Premenstrual and menstrual ischemia and the myth of endometrial arteriovenous
anastomoses," *American Journal of Anatomy* 98 (1956), pp. 69–95.

"The phases of the menstrual cycle and their interpretation in terms of the pregnancy
cycle," *American Journal of Obstetrics and Gynecology* 74 (1957), pp. 931–1055.

Bartelmez, George W. with C. Cuthbertson, "Histologic studies of menstruating
mucus membrane of the human uterus," *Carnegie Contributions to Embryology* 24
(No. 142) (1933), pp. 141–86.

Bartelmez, George W., George W. Corner, and Carl G. Hartman, "Cyclic changes in
the endometrium of the rhesus monkey," *Carnegie Contributions to Embryology*
34 (1951), pp. 99–144.

Benson, Keith R., "American morphology in the late nineteenth century: the Biology
Department at Johns Hopkins University," *Journal of the History of Biology* 18 (2)
(1985), pp. 163–205.

Biggers, John D., "Introduction of the first Carl G. Hartman Lecture," *Biology of
Reproduction* 2 (1970), pp. 1–4.

Bishop, David W., "Biology of spermatozoa," in William C. Young (ed.), *Sex and Internal Secretions*, 3rd edn. (Baltimore, MD: Williams and Wilkins, 1961), pp. 707–
796.

Blake, John B., "Anatomy," in Ronald L. Numbers (ed.), *The Education of American
Physicians* (Berkeley, CA: University of California Press, 1980), pp. 29–47.

Bodian, David, "George William Bartelmez, 1885–1967," *Biographical Memoirs of the
National Academy of Science, USA*. XLIII (New York: Columbia University Press,
1973), pp. 1–26.

Boving, Bent, "Biomechanics of implantation," in R. J. Blandau (ed.), *The Biology of the Blastocyst* (Chicago, IL: University of Chicago Press, 1971), pp. 423–42.

"Observing inside the living uterus," in J. C. Daniel, Jr. (ed.), *Methods in Mammalian Embryology* (San Francisco, CA: W.H. Freeman, Co., 1971).

Burns, Robert Kyle, "The role of hormones in the differentiation of sex," in William C. Young (ed.), *Sex and Internal Secretions*, 3rd edn. (Baltimore, MD: Williams and Wilkins, 1961), pp. 76–160.

Casper, Monica, *The Making of the Unborn Patient: A Social Anatomy of Fetal Surgery* (New Brunswick, NJ: Rutgers University Press, 1998).

Cattell, Jacques (ed.), *American Men of Science II: Biological Sciences* (Lancaster, PA: The Science Press, and New York: R.R. Bowker Company, 1955).

Clarke, Adele E., "Research materials and reproductive science in the United States, 1910–1940," in Gerald L. Geison (ed.), *Physiology in the American Context, 1850–1940* (Bethesda, MD: American Physiological Society, 1987), pp. 323–50. Reprinted in S. Leigh Star (ed.), *Ecologies of Knowledge: New Directions in Sociology of Science and Technology* (Albany, NY: SUNY Press, 1995), pp. 183–219.

"Women's health over the life cycle" in Rima Apple (ed.), *The History of Women, Health and Medicine in America: An Encyclopedic Handbook* (New York: Garland Press, 1990), pp. 3–39.

"Controversy and the development of American reproductive sciences." *Social Problems* 37(1) (1990), pp. 18–37. Reprinted in Andrea Tone (ed.), *Controlling Reproduction: An American History* (Wilmington, DE: Scholarly Resources Inc., 1997).

"Embryology and the development of American reproductive sciences, 1910–1945," in Ronald Rainger, Keith Benson, and Jane Maienschein (eds.), *The American Expansion of Biology* (New Brunswick, NJ: Rutgers University Press, 1991), pp. 107–32.

"Money, sex and legitimacy at Chicago, 1900–1940: Lillie's center of reproductive biology," Special Issue on Biology at the University of Chicago, *Perspectives on Science* 1(3) (1993), pp. 367–415.

"Disciplining Reproduction: Modernity, American Life Sciences, and the "Problems of Sex" (Berkeley, CA: University of California Press, 1998).

"Maverick reproductive scientists and the production of contraceptives *c.* 1915–2000," in Anne Saetnan, Nelly Oudshoorn, and Marta Kirejczyk (eds.), *Bodies of Technology: Women's Involvement with Reproductive Medicine* (Columbus, OH: Ohio State University Press, 2000), pp. 37–89.

Corner, George W., "Ovulation and menstruation in macacus rhesus," *Carnegie Contributions to Embryology* 75 (1923), pp. 73–110.

"The nature of the menstrual cycle," *Medicine* 12 (1933), pp. 61–82.

"Our knowledge of the menstrual cycle, 1910–1950," Fourth Annual Addison Lecture Delivered at Guy's Hospital, London, 13 July 1950, *The Lancet* (28 April 1951), pp. 919–23.

Anatomist at Large: An Autobiography and Selected Essays (New York: Basic Books, 1958).

"The past of anatomy in the United States," *Anatomical Record* 137 (1960), pp. 179–82.

"Foreword," in William C. Young (ed.), *Sex and Internal Secretions*, third edition (Baltimore: Williams and Wilkins, 1961), pp. ix–xii.

"The generality and the particularity of man," in Caryl P. Haskins (ed.), *The Search for Understanding: Selected Writings of the Carnegie Institution* (Cambridge, MA: Massachusetts Institute of Technology Press, 1967), pp. 211–34.

The Seven Ages of a Medical Scientist: An Autobiography (Philadelphia, PA: University of Pennsylvania Press, 1981).

Corner, George W. and Willard M. Allen, "Physiology of the corpus luteum 2. Production of a special uterine reaction (progestational proliferation) by extracts of the corpus luteum," *American Journal of Physiology* 88 (1929), pp. 340–6.

"Physiology of the corpus luteum 3. Normal growth and implantation of embryos after very early ablation of the ovaries, under the influence of extracts of the corpus luteum," *American Journal of Physiology* 88 (1929), pp. 340–6.

Csapo, A. I, "Defence mechanism of pregnancy," in G. E. W. Wolstenholme and Margaret P. Cameron (eds.), *Progesterone and the Defence Mechanism of Pregnancy*, Ciba Study Group No. 9 in honor of George W. Corner (Boston, MA: Little Brown, 1961), pp. 3–26.

"The 'seesaw' theory of the regulatory mechanism of pregnancy," *American Journal of Obstetrics and Gynecology* 121(4) (1975), pp. 578–81.

Duden, Barbara, *Disembodying Women: Perspectives on Pregnancy and the Unborn*, translated by Lee Hoinacki (Cambridge, MA: Harvard University Press, 1993).

Ebert, James D., "Department of Embryology report – introduction," Carnegie Institution of Washington *Year Book 75* (Washington, DC: Carnegie Institution of Washington, 1975–6).

Eskes, T. K. A. B., "Abruptio placentae – a "classic" dedicated to Elizabeth Ramsey," *European Journal of Obstetrics, Gynecology, and Reproductive Biology* 75(1) (1997), pp. 63–70.

Fletcher, Suzanne W., Robert H. Fletcher, and M. Andrew Greganti, "Clinical research trends in general medical journals, 1946–1976," in Edward B. Roberts, Robert I. Levy, Stan N. Finkelstein, Jay Moscowitz, and Edward J. Sondik (eds.), *Biomedical Innovation* (Cambridge, MA: MIT Press, 1981), pp. 284–300.

Gilbert, Scott F., "The embryological origins of the gene theory," *Journal of the History of Biology* 11 (1978), pp. 307–51.

Gillespie, E. C., E. M. Ramsey, and S. R. M. Reynolds, "The pattern of uterine growth during pregnancy in monkeys as shown in an x-ray study," *American Journal of Obstetrics and Gynecology* 58 (1949), pp. 758–64.

Greep, Roy O., M. A. Koblinsky, and F. S. Jaffe, *Reproduction and Human Welfare: A Challenge to Research* (Boston, MA: MIT for the Ford Foundation, 1976).

Greep, Roy O. and Marjorie A. Koblinsky (eds.), *Frontiers in Reproduction and Fertility Control: A Review of the Reproductive Sciences and Contraceptive Development* (Boston: MIT for the Ford Foundation, 1977).

Gruhn, John G. and Ralph R. Kazer, *Hormonal Regulation of the Menstrual Cycle: The Evolution of Concepts* (New York: Plenum Medical Book Company, 1989).

Hahn, R., "Division of labor: obstetricians, women, and society in *Williams' Obstetrics*, 1903–1985," *Medical Anthropology Quarterly* (NS) 1(3) (1987), pp. 256–82.

Hall, Diana Long, "Sex, fertility and taboo: The committee for research on problems of sex, 1920–1940," unpublished manuscript, provided by the author; presented at Workshop on Historical Perspectives on the Scientific Study of Fertility in the United States, American Academy of Arts and Sciences, 1978.

Harlow, Sioban, "Function and dysfunction: an historical critique of the literature on menstruation and work," *Health Care for Women International* 7 (1986), pp. 39–50.

Hartman, Carl G., "Observations on the viability of the mammalian ovum," *American Journal of Obstetrics and Gynecology* 7 (1924), pp. 40–3.

"Bimanual rectal palpation as applied to the female rhesus monkey," *Anatomical Record* 45 (1930), p. 263.

"On the relative sterility of the adolescent organism," *Science* 74 (1931), pp. 226–7.

"Studies in the reproduction of the monkey, *macacus (Pithecus) rhesus* with special reference to menstruation and pregnancy," *Carnegie Contributions to Embryology*, no. 134 (Washington, DC: Carnegie Institution of Washington, 1932), pp. 1–161.

"Ovulation and the transport and viability of ova and sperm in the female genital tract," in Edgar Allen (ed.), *Sex and Internal Secretions* (Baltimore, MD: Williams and Wilkins, 1932), pp. 647–732.

"Catholic advice on the safe period," *Birth Control Review* 17 (May, 1933), pp. 117–19.

Time of Ovulation in Women: A Study on the Fertile Period in the Menstrual Cycle (Baltimore, MD: Williams and Wilkins, 1936).

"Facts and fallacies of the safe period," *Journal of Contraception* 2 (1937), pp. 51–61.

"Studies on reproduction in the monkey and their bearing in gynecology and anthropology," *Endocrinology* 25 (1939), pp. 670–82.

"The scientific achievements of George Washington Corner, M.D.," *American Journal of Anatomy* 98(1) (1956), pp. 5–19.

"How do sperms get into the uterus?" *Fertility and Sterility* 8 (1957), pp. 403–27.

"Annotated list of published reports on clinical trials with contraceptives," *Fertility and Sterility* 10(2) (1959), pp. 177–89.

"The 1960 Ayerst lecture: a half century of research in reproductive physiology," *Fertility and Sterility* 12 (1961), pp. 1–19. Reprinted in Rudolph P. Vollman (ed.), *Fifty Years of Research on Mammalian Reproduction: A Bibliography of the Scientific Publications of Carl G. Hartman* (Washington, DC: USDHEW, Public Health Service Publication No. 1281, 1965).

Science and the Safe Period: A Compendium of Human Reproduction (Baltimore: Williams and Wilkins, 1962).

"Research should spell FUN," in G. W. Duncan, R. J. Ericsson, and R. G. Zimbelman (eds.), *Capacitation of Spermatozoa and Endocrine Control of Spermatogenesis* (Oxford, UK: Blackwell Scientific Publications, 1967), pp. 1–10.

King, Jessie L., "Concerning the periodic cardiovascular and temperature variations in women," *American Journal of Physiology* 34 (1914), pp. 202–19.

"Possible periodic variations in the extent of the knee-jerk in women," *American Journal of Physiology* 42 (1916/17), p. 607.

"Possible periodic variations in the extent of the knee-jerk in women," *American Journal of Physiology* 47 (1918/19), pp. 404–9.

"Menstrual records and vaginal smears in a selected group of normal women," *Carnegie Contributions to Embryology* 95 (1926), pp. 79–93.

"Menstrual intervals," *American Journal of Obstetrics and Gynecology* 25(4) (1933), pp. 583–90.

Kohler, Robert E., *Partners in Science: Foundations and Natural Scientists, 1900–1945* (Chicago, IL: The University of Chicago Press, 1991).

Leonardo, Richard A., *History of Gynecology* (New York: Froben, 1944).

Long, Diana E., "Physiological identity of American sex researchers between the two world wars," in Gerald L. Geison (ed.), *Physiology in the American Context, 1850–1940* (Bethesda, MD: American Physiological Society, 1987), pp. 263–78.

Longo, Lawrence D., "Obstetrics and gynecology," in Ronald L. Numbers (ed.), *The Education of American Physicians* (Berkeley, CA: University of California Press, 1980), pp. 205–25.

"John Whittridge Williams and academic obstetrics in America," *Transactions and Studies of the College of Physicians of Philadelphia* Series V, III (4) (1981), pp. 221–54.

Longo, Lawrence D. and G. Meschia, "Elizabeth M. Ramsey and the evolution of ideas of uteroplacental blood flow and placental gas exchange," *European Journal of Obstetrics, Gynecology and Reproductive Biology* 90(2) (2000), pp. 129–33.

Maienschein, Jane "Epistemic styles in German and American Embryology," *Science in Context* 4(2) (1991), pp. 407–27.

Transforming Traditions in American Biology, 1880–1915 (Baltimore, MD: Johns Hopkins University Press, 1991).

Markee, J. Eldridge, "Menstruation in intraocular endometrial transplants in the rhesus monkey," *Carnegie Contributions to Embryology* 177 (1940), pp. 220–308.

Mitman, Gregg, Adele Clarke, and Jane Maienschein (guest eds.), "Introduction to Special Issue on Biology at the University of Chicago, c. 1891–1950," *Perspectives on Science* 1(3) (1993), pp. 359–66.

Moses, Bessie L., "Dr. Jessie L. King," *Goucher Alumnae Quarterly* XXV (4 August 1947), p. 17.

O'Rahilly, Ronan, "One hundred years of human embryology," *Issues and Reviews in Teratology* 4 (1988), pp. 81–128.

Parkes, A. S., "Prospect and retrospect in the physiology of reproduction," *British Medical Journal* (14 July 1962), pp. 71–5.

"The rise of reproductive physiology, 1926–1940: The Dale lecture for 1965," *Endocrinology* (*Proceedings of the Society*) (1966), pp. xx–xxxii.

Pauly, Philip J., "The appearance of academic biology in late nineteenth century America," *Journal of the History of Biology* 17 (3) (1984), pp. 369–97.

Pearl, Raymond, "Contraception and fertility in 2000 women," *Human Biology* 4 (1932), pp. 363–407.

Petchesky, Rosalind Pollack, "Fetal images: the power of visual culture in the politics of reproduction," *Feminist Studies* 13 (1987), pp. 263–92.

Pierpoint, Raymond, *Report of the Fifth International Neo-Malthusian and Birth Control Conference* (London: William Heinemann, 1922).

Porter, D. G. and C. A. Finn, "The biology of the uterus," in Roy O. Greep and Marjorie A. Koblinsky (eds.), *Frontiers in Reproduction and Fertility Control: A Review of the Reproductive Sciences and Contraceptive Development* (Boston, MA: MIT for the Ford Foundation, 1977), pp. 146–56.

Raacke, I. D., "Herbert Mclean Evans (1882–1971): a biographical sketch," *Journal of Nutrition* 113(4) (1983), pp. 928–43.

Ramsey, Elizabeth M., "The Lockyer embryo: an early human embryo in situ," *Carnegie Contributions to Embryology* 26 (1937), pp. 99–120.

"The Yale embryo," *Carnegie Contributions to Embryology* 27 (1938), pp. 67–84.

"The Carnegie monkey colony and the placental circulation of primates," Carnegie Institution of Washington *Year Book* 70 (1970–1), pp. 84–93.

The Placenta: Human and Animal (New York: Praeger, 1982).

Ramsey, Elizabeth M., G. W. Corner, Jr., M. W. Donner, and H. Stran, "Radioangiographic studies of circulation in the maternal placenta of the rhesus monkey: preliminary report," *Proceedings of the National Academy of Science USA* 46 (1960), pp. 1003–8.

Ramsey, Elizabeth M., G. W. Corner, Jr., and M. W. Donner, "Cineradiographic visualization of venous drainage of the primate placenta in vivo," *Science* 141 (1963), pp. 909–10.

"Serial and cineradiographic visualization of maternal circulation in the primate (hemochrial) placenta," *American Journal of Obstetrics and Gynecology* 86 (1963), pp. 213–25.

Ramsey, Elizabeth M. and M. W. Donner, *Placental Vasculature and Circulation* (Stuttgart: Georg Thieme, 1980).

Reed, James, *The Birth Control Movement and American Society: From Private Vice to Public Virtue,* 2nd edn. (Princeton, NJ: Princeton University Press, 1983).

Reynolds, Samuel R. M., "The nature of uterine contractility: a survey of recent trends," *Physiological Review* 17(2) (1937), pp. 304–34.

The Physiology of the Uterus (New York: Paul B. Hoeber Pub., 1949).

"Gestation mechanisms," in Otto V. St. Whitelock (ed.), *Special Issue: The Uterus, Annals of the New York Academy of Sciences* 75, Art. 2 (1959), pp. 691–9.

The umbilical cord," in Caryl P. Haskins (ed.), *The Search for Understanding: Selected Writings of the Carnegie Institution* (Cambridge, MA: Massachusetts Institute of Technology Press, 1967), pp. 181–90.

Reynolds, Edward and Donald Macomber, *Fertility and Sterility in Human Marriages* (Philadelphia, PA: W.B. Saunders, 1924).

Sabin, Florence R., *Franklin Paine Mall* (Baltimore, MD: Johns Hopkins University Press, 1934).

Sanger, Margaret (ed.), *Biological and Medical Aspects of Contraception, American Conference on Birth Control and National Recovery.* (Washington, DC: National Committee on Federal Legislation for Birth Control, Inc., 1934).

Spradling, Allan C., "The Carnegie Institution of Washington, Department of Embryology," *Molecular Medicine* 3(7) (1997), pp. 417–19.

Sprague, James M., "Louis Barkhouse Flexner, January 7, 1902–March 29, 1996," *Biographical Memoirs of the National Academy of Sciences, USA* 73 (1998), pp. 3–17.

Stevens, Rosemary, *American Medicine and the Public Interest* (New Haven, CT: Yale University Press, 1971).

Stewart, T. Dale, "Adolph Hans Schultz," *Biographical Memoirs of the National Academy of Sciences, USA* 54 (1983), pp. 325–49.

Viterbo, Paula, *Counting the Days: A History of Natural Birth Control in America* (Cambridge, MA: Harvard University Press, forthcoming).

Vollman, Rudolph P., Preface in "Carl Hartman's contributions to the physiology of reproduction," pp. v–vii in Rudolph P. Vollman (ed.), *Fifty Years of Research on Mammalian Reproduction: A Bibliography of the Scientific Publications of Carl G. Hartman* (Washington, DC: USDHEW, Public Health Service Publication no. 1281, 1965).

Young, William C. (ed.), *Sex and Internal Secretions,* 3rd edn. (Baltimore, MD: Williams and Wilkins, 1961).

5

THE LEWIS FILMS: TISSUE CULTURE AND "LIVING ANATOMY," 1919–1940

HANNAH LANDECKER

Department of Anthropology, Rice University, Houston

When skin and eye cup you brought close,
Harmoniously, a lens arose.
Yet, dorsal lip you dislocate,
Keeps up its will to gastrulate.
While gel layer and plasm inside
Cooperate to make cells glide,
Their eager tips, forever thirsting,
Pinocytose till well-nigh bursting . . .
            Paul Weiss, A Poem in Honor of Dr. Lewis[1]

This chapter will focus on the work of Warren Harmon Lewis and Margaret Reed Lewis in the observation of the behavior of living cells. From the current vantage point of the early twenty-first century, in which cell biology and cell biotechnology are ascendant, it is necessary to emphasize that not only did they observe cells, they also participated in the establishment of the very idea of the existence and importance of cellular behavior in the first place. By helping develop and use the first experimental systems for doing cellular physiology of the somatic cells of complex organisms, the Lewises participated in the demonstration that somatic cells were autonomous entities whose behavior over time (motility, ingestion, division, change of shape, transformation) was intimately related to biological questions of the whole organism: development, infection, immunity, physiology, cancer.

As historian and embryologist Jane Oppenheimer dryly reminded an audience in 1970, embryologists have not always treated cells as being the protagonists they are today in narratives of development: "Cells are very popular in 1970; embryologists who liked organizers in 1925 were less concerned with cells *qua* cells."[2] The point of her talk in this instance was to highlight Johannes Holtfreter's experiments with isolated cells and embryonic parts *in vitro* in the 1930s through 1950s. Holtfreter aside, the historical question posed stands: what influenced embryologists to pay "attention to cells as organized individuals," and to think in terms of cells rather than supracellular agencies?

One answer lies not in the person of this embryologist or that, but in the introduction of the *in vitro* techniques of cell and tissue culture that many began to use over the twentieth century for the investigation of the life of cells. The Lewises contributed a great deal of foundational work in this venture. Again from the current vantage point of tissue culture as a ubiquitous laboratory technique, it is necessary to emphasize that before the availability of antibiotics, the use of fume hoods, the development of standardized growth mediums, and standardized glassware, and before biologists could open a catalog and order a cell line, growing living somatic cells of complex organisms outside the body was a technically challenging venture whose possibilities and limitations were simply unknown. It was in systematically exploring and establishing some of the basic parameters of tissue culture, and then training generations of other biologists in these techniques, that the Lewises had their greatest impact on twentieth century embryology and cell biology.

In 1964, when Warren Harmon Lewis died at the age of 94, his wife and long-time scientific collaborator, Margaret Reed Lewis, sent out copies of her husband's bibliography in lieu of cards. That a completed publication list seemed the fitting mark of the end of a life shows only the absolute inextricability of biography and biology for the Lewises, who had worked together daily in the laboratory from 1910 until a few weeks before Warren Lewis' death. This article necessarily focuses on only some aspects of this long and productive career, by examining their work at the Carnegie Institution of Washington (CIW) Department of Embryology from 1919 to 1940. During this time they concentrated on experiments with tissue culture, the technique of growing living cells *in vitro*, outside of the bodies of complex organisms.

Because it would be impossible to recount all that they did during these years, this chapter will work to highlight a strong thematic link that runs through this work: the attention the Lewises called to the cell as a dynamic agent of change, which served as the basic legitimation for the scientific significance of the then-novel experimental technique of tissue culture. After all, no one else would have been interested in using or developing the method if there weren't new and interesting things to see and do with cells; the demonstration of experimental possibilities previously foreclosed by the technical limitations of older methods of histology and anatomy was key to the eventual uptake of this technique by a much broader community. Whether talking of muscle contraction, mitochondria, transformation of blood cell types, cancerous cells, or processes of infection, their emphasis was on the cell's own integral and particular abilities to move, reproduce, transform, grow, ingest, or otherwise change. Thus, this chapter will begin with a brief consideration of how the Lewises became involved in the early development of tissue culture, and trace their concentration on the dynamism of the cell through several areas of investigation.

A methodological note is necessary here. Many of this chapter's conclusions are drawn from paying close attention to the way the Lewises wrote about cells. A comprehensive account of the content of all their research would be overly long and tedious; such a list would also be misleading about the coherency of this body of work, because it would appear to be a list of separate observations on muscle contraction, the interactions of bacteria and somatic cells, mitochondria, white blood cells, cancer, and pinocytosis. It would also necessitate treating the two scientists separately, since although they published a great deal together, they also published separately on different topics.[3] While some of these publications have turned out to be more memorable than others, I have chosen instead to focus on what bound their work together, and its significance not as a set of results but as a distinctive way of looking at, manipulating, and thinking about cells. As such it is a case study in the larger phenomenon of how living cells became the fundamental laboratory tools that they are today.

A close reading of how the Lewises narrated cellular life in their scientific papers yields a synthetic picture of their increasing realization of – and respect for – the cell as a primary agent, the key player in life processes, with intrinsic powers to move and ingest and change and react, phenomena which would have to be described if higher-order processes of development and disease were ever to be understood. This synthetic picture also gives a better indication of why the directors of the Department of Embryology stood behind this research for more than twenty years as important and relevant to embryology, when in some aspects it might have appeared more relevant to pathology or general physiology than to human embryology. That is, directors like George Streeter were able to recognize common questions of change and transformation for cells and embryos in the way the Lewises framed their experiments and findings. As he wrote to John C. Merriam, President of the CIW:

> Tissue culture provides another means of study of cell structure and cell physiology and thereby provides the foundation on which embryology must rest. To know to what extent the single cell possesses a permanent individuality, in its form and behavior, is a prerequisite to the understanding of their cooperative behavior en masse or in their integration into a tissue.[4]

Therefore, understanding the vision of the cell as a dynamic component of the developing body behind all the Lewis' work is also vital to understanding these scientists' institutional place within the Department of Embryology.

Observing live somatic cells outside of the body was a new field of research, and thus, "for these pioneers, practically every observation yielded pay-dirt";

these prolific results were used not only to build knowledge of the cell, but to work constantly on the techniques with which to grow and observe them better.[5] The observation of cells was built directly back into the techniques of their maintenance outside of the body. Central to this work was an emphasis on the visibility of the cell afforded by tissue culture; the ability to watch and narrate life processes. In the late 1920s, the use of time-lapse microcinematography for the close observation and behavioral analysis of living cells on film became a centerpiece of the laboratory. Many cellular changes happen at a time-scale too slow for normal human perception; taking an image at regular intervals and then projecting the film at normal speed accelerated slow change and made it accessible for repeated observation. Cinematography thus became both the method and expression of investigation of living cells, particularly for Warren Lewis, and came to typify the Lewis' campaign to practice "living anatomy" via tissue culture. It is a central irony of Warren Lewis' life that for twenty-four years he was the editor of *Gray's Anatomy*, the central tome of classical anatomical depiction, bringing out editions 20 through 24, while simultaneously working tirelessly to break with exactly that kind of morphology, dissection, and static illustration of dead tissues, by developing cinematic methods for observing live tissues.[6]

The cultivation of cells

Warren Harmon Lewis began his career studying anatomy and embryology, first as a medical student and then as a professor at the Medical School of the Johns Hopkins University, joining the Department of Anatomy headed by Franklin P. Mall in 1900.[7] After a period of work in Germany in the laboratory of Moritz Nussbaum at Bonn, he returned to Baltimore and began research on the embryonic development of the eye. Through a series of delicate operations performed under the dissecting microscope, he transplanted embryonic skin into the eye cup, where the skin was converted into a lens, and transplanted optic cup under ectoderm elsewhere on the body, which induced lens formation where it would not normally take place.[8] He also took amphibian embryos at the late gastrula stage and transplanted fragments of the blastopore to other parts of the embryo, experiments that Hans Spemann would repeat some years later but interpret very differently, indeed to form the basis of Spemann's general organizer theory.[9] These experiments in embryonic development, undertaken in direct reference to what other investigators such as Ross Harrison, Nussbaum, and Hans Spemann were doing in experimental embryology in Germany and the USA, seemed the beginning of a career spent investigating induction and development, writing reference works on the human developmental anatomy, and teaching embryology and anatomy.

This solid academic career swerved unexpectedly due to two factors: Lewis took up the new field of tissue culture, and joined the CIW Department of Embryology in 1919. The combination of an utterly new technique and field of study, and a research position removed from university teaching resulted in a life spent observing cells. The nature of Lewis' work and publications changed quite abruptly when he married Margaret Reed in 1910 and they turned their collective attention on to the exciting new possibilities of cell cultivation.[10] In the period 1907–10, great interest was spurred in the possibility of growing isolated fragments or cells of complex organisms outside of the body by a series of experiments done by the embryologist Ross Harrison, who was Lewis' colleague at Johns Hopkins until moving to Yale in 1907.

Harrison, seeking resolution to the question of the origin of the nerve fiber, isolated small fragments of embryonic amphibian nerve tissue. Suspended in a drop of lymph from a cover slip over the depression in a hollow slide, these hanging drops could be observed under the microscope. Harrison found that rigorous aseptic practice created conditions in which the cells in the fragment could remain alive for some weeks, and begin to move and differentiate despite their isolation from the rest of the embryonic body.[11] His immediate aim was to remove the cells from the confusing complexity of the body, in order to determine whether the nerve fiber arose from a single cell or from the fusing of many cells; he needed the cells to stay alive long enough to watch the process of nerve outgrowth as it happened.[12] Many observers of the experiment – including the Lewises – were as moved by the method as by the results; the surprising but manifest ability of somatic cells to live and move and differentiate outside of the body, where they were isolated, easy to manipulate, and amenable to continuous direct observation, suggested enormous experimental potential.

Among those observers was the surgeon Alexis Carrel at the Rockefeller Institute for Medical Research, who was interested in getting at the biological mechanisms of wound healing and tissue regeneration. At the urging of Simon Flexner, the institute's director, Carrel sent his assistant Montrose Burrows (also a Johns Hopkins graduate) to Harrison's lab to learn the hanging drop technique. Carrel and Burrows rapidly adopted and appropriated the method to Carrel's experimental program, substituting mammalian tissues and blood plasma for the amphibian tissue and lymph developed by Harrison.[13] Carrel was interested in the phenomena of ongoing life and growth, not embryonic differentiation, and quickly altered the time-frame of the method from a one-time culture with a life-span of days or weeks to a system of continuous culture, whereby new cultures of cells could be made by taking a fragment of the old culture, and immersing it in fresh nutrient medium.

Margaret Reed received her A.B. degree in physiology from Goucher College in 1901, and went on to Bryn Mawr, working with Thomas Hunt

Morgan, who she then followed to Columbia. After meeting Warren Lewis at Woods' Hole, she resigned from teaching physiology and biology at Barnard upon her marriage, and became an unpaid contributor in his lab at Johns Hopkins. This was a relatively typical pattern for women pursuing laboratory science early in the twentieth century.[14] Like her husband, she had spent time in Germany, but with Max Hartmann in Berlin doing experimental cytology in 1908. Reed had while working in Berlin with Rhoda Erdmann attempted to grow guineau pig bone marrow tissue on the agar used to cultivate amoebae, and had observed "a membrane-like growth with mitotic figures" on the surface of the agar.[15] With these experiments in mind, and inspired by Harrison's success with nerve tissue, they attempted to culture embryonic pig bone marrow in blood plasma, but without conclusive results until Carrel and Burrows published their adaptation of Harrison's technique.

Once they were able to attain growth of tissues *in vitro*, the tone of the Lewises' first publications on tissue culture and its potential as an experimental system was one of barely restrained excitement. They wrote of the "very wonderful growths from the various organs of the embryo" that they were able to achieve, observing "The cells are active, sending out and retracting their pseudopodia."[16] For the Lewises, what Harrison, Carrel, and Burrows had done marked no less than "the beginning of a new epoch in experimental anatomy."[17] However, the work was hardly begun. Everything was possible, and nothing was decided. The Lewises began immediately to experiment with different media, recognizing the potential for controlled experiment in studying tissues in "media all the constituents of which are known and the reaction of the cells to different substances can be more definitely determined."[18] They found to their surprise that not only would cells grown in plasma and lymph, they also did so (even if for a limited period of time) in bouillon, agar, nutrient media and even salt solutions. It was with some astonishment that they reported on the apparent resilience and plasticity of cells extracted from the body: "It is quite remarkable that cells will grow in such widely different solutions . . ."[19] Their Locke–Lewis solution (salt solution supplemented with bouillon and dextrose), published in 1910, became a mainstay of tissue culture practice, although a fully defined medium which would support the unlimited growth of cells remained elusive until half a century later.

Over the next decade, as they made the transition into the Department of Embryology, the Lewises simultaneously explored how to grow cells, and what one could see having grown them; their publications were both explications of, and encomiums to, the new methods and what they rendered visible. They worked primarily with small pieces of tissue suspended in a drop of medium over a hollow slide: "In a hanging drop of medium the growth adheres closely to the cover slip; the cells are so thin that, if desired, their most minute structure can be observed by means of the oil immersion lens and high oculars."[20] This isolation of cells from the complexity of the body worked both to render them visible, and to sort out what was an intracellular rather than an intercellular phenomenon. Just as Harrison had shown the autonomous growth of the nerve fiber out from a single cell, the Lewises, following on early experiments of Montrose Burrows, investigated the question of whether muscle contraction too could be an inherent quality of single cells, concluding that "the power of rhythmical contraction exists in the smooth muscle cells of the cultures of the chick amnion as an *inherent property* of the protoplasm of the smooth muscle cells and may be exhibited either by a bundle of cells, by an individual cell or by a single cell".[21] While it may seem redundant to observe that either an individual cell or a single cell could exhibit contraction, it should be remembered that at this time the source and cause of muscle contraction was a highly contentious issue, and therefore it would be significant to readers whether a cell was contracting as an individual but still in contact with other cells, or contracting even though single and unattached to other cells.[22] Like many phenomena in biology at this time, such as infections and toxins, it was unclear at what level the mechanism of action was working: tissue? organ? body? cell? These mechanisms were difficult to observe in process in the body; tissue culture and its glass-enclosed lives of cells were thus seen by its proponents as unprecedented windows (literally) onto the internal workings of the body, both embryonic and adult. George Streeter, the Director of the Department of Embryology responsible for recruiting the Lewises, was a strong supporter of their work and its relevance to embryology for this reason.

Again, from a contemporary perspective, it is useful to highlight the fact that the somatic cells from complex organisms were at this time *not* thought of as highly autonomous entities with their own behavior, and it was part of the promise and fascination of the technique that scientists could see these sometimes startling behaviors. There is evidence that many who tried their hand at tissue culture for the first time also chose heart tissue to begin with. The sight through the microscope of the pulsating tissue was apparently singularly affecting. Henry Field Smyth reported in the *Journal of the American Medical Association* (*JAMA*) in 1914 that he had attained "most satisfactory growths from seven-day to eleven-day hearts," which "pulsate so violently that they are apt to tear loose from their plasma supports."[23] The cytologist William Seifriz wrote to Albert Ebeling (who worked with Carrel at the Rockefeller) in 1933, noting that he had over the years used heart tissue to teach the techniques to students, because "the continued beating of a heart fragment is one of the things that the student always enthuses over." In fact, he wrote, "I must admit the same naïve feeling myself every time I see it." Seeming defensive that Ebeling would think him "altogether too childish," he hastened to add that he thought that other scientists felt the same way: "if I remember correctly Lewis was no less enthusiastic when he told me of the pulsations which he had observed in a single cell."[24]

Even though tissue culture in this period was done with fragments of tissue (single cell culture was not shown to be possible until the 1940s), the focus fell on the living cell, and what was seen seemed a sharp contrast to the fix and stain methods of histology previously used to study cells and tissues. Writing of their observations of mitochondria, the Lewises again emphasized the ability to see not just new things but to see more clearly the limitations and artifacts of other methods:

> Tissue cultures afford a new and somewhat different method from that usually employed for the study of many cell structures. It enables one to compare the living with the fixed material. In fact, one can study the same cell while living, during the process of fixation, and later as a stained permanent preparation. It also enables one to follow the changes which take place in the living cell from minute to minute.[25]

The changes from minute to minute introduced a new conception of not just the cell but its internal constituents as highly dynamic, highly variable entities.

> The mitochondria are almost never at rest, but are continually changing their position and also their shape. The changes in shape are truly remarkable not only in the great variety of forms, but also in the rapidity with which they change from one form to another. A single mitochondrium may bend back and forth with a somewhat undulatory movement or thicken at one end and thin out at the other with an appearance almost like that of pulsation, repeating this process many times. Again, a single mitochondrium sometimes twists and turns rapidly as though attached at one end, like the lashing of a flagellum, then suddenly moves off to another position in the cytoplasm.[26]

Building a science of cellular behavior

In 1912, Ross Harrison had commented that the most promising application of the new tissue culture method that he had discovered was that of obtaining "direct knowledge" of cellular activity. Tissue culture had opened up the possibility of extending the notion of "behavior" to include all cells, not just the limited examples of single-celled animals, sperm, the early stages of marine embryos, and some blood cells, the kinds of cells most often observed in the living state prior to tissue culture.

> One of the most striking and interesting things described by Max Schultze in his study of protoplasm was the fact that certain cells of the body, the leucocytes, exhibit activities of movement quite similar to those of some of the most primitive animals . . . But unfortunately we possess even now, fifty years after this work of Schultze's, but scant direct knowledge of cellular activity. The microscope has revealed to us much of the structure of cells and organs, and physiological experiments have taught us the intricacies of the mechanisms of movement, secretion and sensation, but as compared with our knowledge of these fields, that of cell behavior is extraordinarily meager. It is in making up this lack in our knowledge that the methods which we are discussing to-day will find their principal application.[27]

However, it was one thing to identify a research opportunity – that of moving from the structure of cells and the physiology of organs, to the physiology and behavior of cells themselves – it was another to come to precise experimental methods for observing and understanding cellular behavior. Many who tried their hand at tissue culture in the very early years quietly dropped it again when it turned out to be quite difficult and labor-intensive. The Lewises, by contrast, continuously worked to improve their methods and to tailor them to specific biological questions. By the time the Lewises formally joined the Department of Embryology in 1919, they had turned almost exclusively to working with tissue culture systems.

In these passages from Margaret Lewis' experiments with the reaction of embryonic chick cells to bacilli, one gets a sense first of the intense physicality of "observation" in this context, and the corresponding effort to give "behavior" a highly specific reality for the reader: "These cells must possess great oxidizing power, for they took in and destroyed many organisms within a short period of time (thirty minutes to two hours). When a single cell was kept under observation, the actual number of organisms destroyed could be determined. One such cell had ten bacilli moving in the cytoplasm at the beginning of the observation and during the two hours it was watched it destroyed these and thirteen more."[28] This passage is absolutely representative; it is clear that the Lewises built their results by spending their time doing things like watching a single cell for hours. Along with this intensity of observation came an awareness, paradoxically, of all they *could not* see; significant things were most evidently happening ("these cells must possess great oxidizing power"), but it was not entirely clear exactly what. Along with the note of respect for the powers of the living cell to ingest foreign bodies, there is an allied note of frustration, of only being able to access the superficial levels of the phenomenon. "The ingestion and destruction of these organisms, which seems such a simple performance as one watches it in tissue-cultures, certainly emphasizes again the remarkable structure of the living cell. The motility and the gradual fading away of the bacilli indicate . . . that intense chemical changes are taking place, even though we cannot see them."[29] George Corner commented in his obituary for Warren Lewis that at this point the Lewises had opened so many different avenues of inquiry that seemed amenable to the tissue culture method (mitochondria, muscle cells, media development, cancer, immunity, and general physiology of the cell) that they began to divide up the labor and work with different concentrations. While, as he put it, it was still as hard to distinguish "between their respective ideas and accomplishments as between two contiguous fibroblasts," Margaret Lewis focused her attention on microbiological problems such as pH changes in cultures and the cultivation of viruses in living cells, while Warren Lewis turned his attention to the question of descriptive cellular morphology and cell mechanics, and to the development of methods of cinematography.[30]

While they did indeed venture into different problems of microbiology and cell morphology and publish separately on these aspects of cellular life, there is an interesting continuity to their continuing exploration of tissue culture as providing a kind of microcosm in which to elucidate nagging questions of reaction and transformation inside the body, both between the body's different components, and between the body and foreign substances, such as infectious bacteria. For Margaret Lewis, this took the form of close observation of the reaction of cultures of chick embryo intestine to typhoid bacilli added to the medium.[31] She observed a very specific form of reaction on the part of the cells, characterized by a "foamy" appearance. Since typhoid bacilli affect the intestines in the course of disease in the body, this was indeed a very specific microcosm of what might be happening in the body. Once again, method and result were simultaneously hailed. Tissue culture could isolate the infectious event in a way specific to what might happen in the complex conditions of the whole body; the cellular reaction made visible thereby showed that infection and disease was fundamentally a *cellular* phenomenon.

Another significant avenue of investigation opened by watching live cells over time was the question of life-cycles of cells, or origin and transformation from one type to another, questions that were particularly contentious in studies of white blood cells and cancerous cells. In the 1920s, the Lewises became deeply interested in this set of inter-related and obstinate problems of classification and identification of bodily components, and of identity and transformation of various blood cell types and cancerous cells. Again, it depended on having a microcosm in which to observe particular changes over time. In this case, it was not the reaction of cells to foreign bodies, but the relation of certain cells to others in a temporal sequence. It was not entirely clear what leucocytes had to do with macrophages, or epithelioid cells with clasmatocytes, or the relationship among the variety of other names and classifications given to the different shapes and types of white blood cells that had been identified. In a passage which typically focused a sharp observational eye on the cells in question, Warren Lewis commented on the "peculiar individualism" of cells in the blood stream, which was related to both their motility and their constant change in shape.

> The main difficulties in attempting to determine the relationships between the cells with which we are dealing are due, first, to their migratory character and to their circulation in the blood stream. This is associated with a peculiar individualism, in that they do not unite with others of their kind to form fixed tissues. It is fortunate for anatomists and pathologists that the other types of body cells tend to cling together in more orderly fashion. A second difficulty is their changing character. Being highly phagocytic cells, they have the ability to ingest and digest various foreign bodies, and thereby undergo astonishing changes in size and appearance.[32]

When these cells were spread on slides and stained and fixed, in the classic histological method, this variety of shapes and sizes could lead to mistakes in classification, since a cell that looked different from another might only be different in shape, not type. Furthermore, this static method made it very hard to identify whether one type of cell was a precursor of another, or a different type altogether. Once again, watching the cells as they moved and transformed seemed to be the key. Lewis, in this Harvey lecture, credited Margaret Lewis with having come up with a way of adapting the hanging drop method to blood cells, and,

> By this simple but effective technique the puzzling problem of the inter-relation between these cells seems to us to be quite conclusively settled. Crucial experiments of this sort far outweigh interpretations based on fixed and stained material, or on living material where the intermixture of various tissues renders the field of observation so complex that the eye cannot follow the changes with great certainty.[33]

These observations contributed to the determination of a developmental series of the different types of blood cells, conclusively showing that some developed from others, and that cells thought to be of discrete types were actually different manifestations of the same type. Thus, monocytes and macrophages were shown to represent different states of the same cell, rather than distinct types of cell. Similarly, cells seeming to occupy specific roles in specific places in the body were shown by Warren Lewis to be a particular manifestation of the same cells found elsewhere in the body, for example, the "Hofbauer cells" identified by anatomists as specific to the human placenta were actually macrophages.[34]

With cancerous cells, the same method of extended observation of cells over time led to opposite conclusions about continuity between cell types, this time between normal cells and cancerous cells. Again, the very material sense of labor, of hours spent watching thousands of cultures, comes through in the conclusion that cancerous cells are permanently altered to a different physiological state from normal cells. "It took long training and persistent effort and the examination of many thousands of cultures to convince myself that the malignant cells are visibly different from normal ones, and also to eliminate the idea that there might exist transition forms between normal and malignant ones. . . ."[35]

Beneath the Lewises' apparently diverse excursions in typhus and cancer, immunity and physiology was the unifying attempt to discern exactly what cells were doing in the processes of infection, reaction, transformation, disease and movement. In sum, by 1940, the Lewises had presented a comprehensive description of the living cell and its components, including the mitochondria, explored the physiological activity of cells such as the locomotion of leucocytes and the contraction of muscle cells, experimented with different media and media pH to optimize growth of cells in culture, cleared up the classification of white cells by watching the formation of macrophages and other cells from mononuclear white cells, described the behavior of macrophages in the inflammatory process, explored the mechanism of bacterial infection and the response by the cells of the immune system, detailed the cytological features of malignant cells and ascertained that they are permanently altered from the normal state, and developed a means to communicate these findings not just with words, but with the phenomena themselves, demonstrated on film. I have chosen to focus on film in the final part of this chapter, because it epitomises the effort made by the Lewises to make the observation of cells over time into a specific instrument of investigation into cellular phenomena.

Cinematography

Those privileged to see the Lewis–Gregory motion pictures of fertilized mammalian eggs, mammalian leukocytes, and mammalian tissue cultures have seen dramatized many familiar deductions from histology. Two leukocytes racing and struggling for possession of a single food particle, with the cytologic reactions of victory and defeat; cytocannibalism with the elimination of a senile brother; rhythmic cycles of cellular activity and cellular sleep; and the ceremonious ballet of the chromosomes heralding the approach of nuclear division. Accompanying these expected phenomena, however, are numerous surprises, reactions unpostulated by conventional histology. A prominent example is the phenomenon of hydrophagocytosis, in which macrophages mechanically drink the surrounding plasma.[36]

For all their successes of getting cells to live in full visibility, the Lewises often expressed frustration at their inability to perceive exactly what was going on. Not only was it difficult to see the very slow or very fast movements of the cells, it was difficult to convey these sights to readers. Their sentences often strain at the limitations of words to describe what they spent so much time carefully observing, as in this description of muscle contraction: "The thickening of the muscle strand took place within the three processes and resulted in a peculiar coiling difficult of description, except in the rather picturesque language used by one observer, who remarked that it resembled a wriggling mass of worms."[37] Or, later, in speaking of observations on cancerous cells, Warren Lewis stated that "To give a word picture of the visible characters of the various malignant cells that would enable you to recognize them is an almost impossible task."[38]

However, by 1936, the Lewises no longer depended on their "word pictures" to put across their results to colleagues, because in the late 1920s, they shifted their experimental technique away from direct observation through the microscope, to observing cells filmed with time-lapse microcinematography. As can be seen from the work detailed above, it was from the very beginning evident that there was much to find out about the behavior of cells; film as a dynamic medium, which did not depend on killing the subject in order to observe it, must have seemed like the perfect experimental tool. Warren Lewis traveled to England in 1927 to visit Ronald Canti at the Strangeways Laboratories in England, with the specific purpose of examining Canti's cinematographic set-up. He, Canti, and Honor Fell then traveled together to the Tenth International Zoological Congress in Budapest, which was that year organized around the theme of tissue culture (Ross Harrison gave a keynote address on "The status and significance of tissue culture"), and Lewis saw first-hand the impact of using film to communicate observations to colleagues.

Canti's "außerordentlich schönen Films" were, according to the editors' note, by popular demand shown on three separate occasions during this congress. Canti began his presentation by saying that film as a technique had been chosen to address problems "which for various reasons could not be solved by direct visual observation": "It was thought that by elaborating a cinema technique it would be possible to obtain records over long periods of time, which could be examined in a much shorter period of time, and further, which could be reexamined on as many occasions as might be required for their interpretation. Again it would be possible to run the film backwards and observe the events in the reverse order of time and thus trace to their origin any changes which might have taken place. In short it was decided that a technique which could afford opportunities such as these would constitute an ideal method for biological investigation."[39] Lewis agreed, and by 1929, was producing films of very high quality, far surpassing those of Canti in focus, steadiness, and clarity (Fig. 5.1). There is no doubt that Lewis had "good hands," as scientists like to say of those who seem to have a particular dexterity at the bench, since it was difficult to control simultaneously the vibration of all the parts of the cinematographic apparatus, which could blur the picture, and all the other parameters such as light (whose heat could easily kill the cells), depth of focus, the degree of acceleration necessary to observe a particular movement or change, and the cultures themselves, which had to be in a state suitable to demonstrate the particular phenomenon in question.

The results of using film were immediately evident, in that the Lewises began to see things they weren't able to see by direct observation. In 1929, Warren Lewis observed a phenomenon of cell behavior on film that he called *pinocytosis*, "drinking by cells," meant to complement the earlier *phagocytosis*, "eating by cells." *Phagocytosis* was coined by Metchnikoff in the 1880s as a term to describe the cellular intake or engulfment of solid particles of matter. Pinocytosis, by contrast, was first seen, described, and coined as a term by watching films of macrophages moving across the cover glass of hanging-drop preparations, away from explants of rat tissue. "The wavy, sheet-like processes or thin membranous pseudopodia of the macrophage series – monocytes, macrophages (clasmocytes), epithelioid cells – are often elaborate and project

Figure 5.1   Warren Lewis projecting a film. (*Philadelphia Inquirer* Magazine, May 11, 1952, p. 16.) Reprinted with permission of the *Philadelphia Inquirer*.

out from the body of the cell as waving sheets. The curious motions which they undergo are much emphasized by motion pictures."[40]

Lewis observed these cells taking in globules of fluid from the surrounding medium, which he presumed contained proteins, water, and various salts. He was terse about the implications of such cell "drinking," saying only that "the importance of this phenomenon in cellular metabolism and the economy of intercellular fluids seems almost self-evident," and he apparently showed the films of pinocytosis around the country for two years prior to publishing the results in written form. Others, however, were more overtly enthusiastic. An editorial in *JAMA* about the films observed that: "Physiologists will find in the Lewis phenomenon suggestion of a new mechanism of cellular nutrition and mechanical filtration or purification of body fluids."[41] This editorialist also noted that the films were available to all for rental or sale.

In an article illustrated by sixteen frames of a film of pinocytosis (Figs. 5.2 and 5.3), Lewis described the spectacle of pinocytosis; again, the emphasis was on continuous movement.

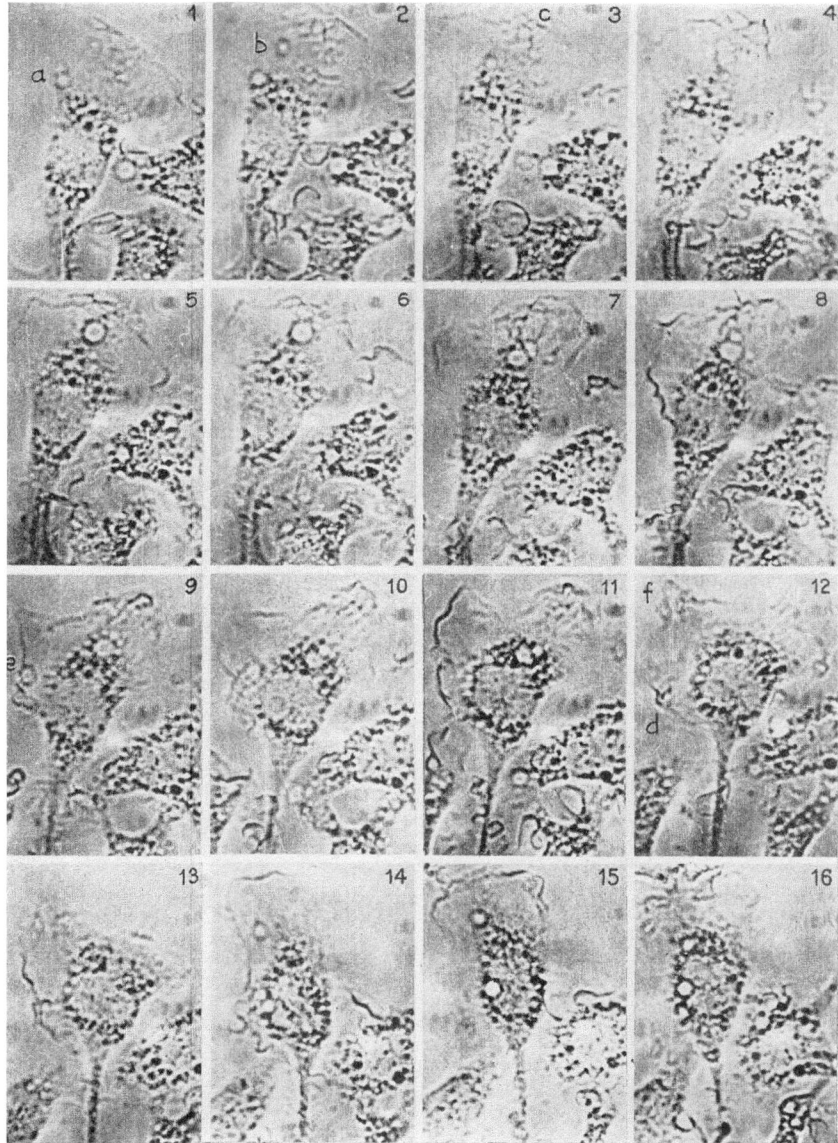

Figure 5.2   Film stills from the film "Pinocytosis." (Lewis, Warren H. Pinocytosis. *The Johns Hopkins Hospital Bulletin* 49 (1929), 26, 27. © The Johns Hopkins University Press. Reprinted with permission.)

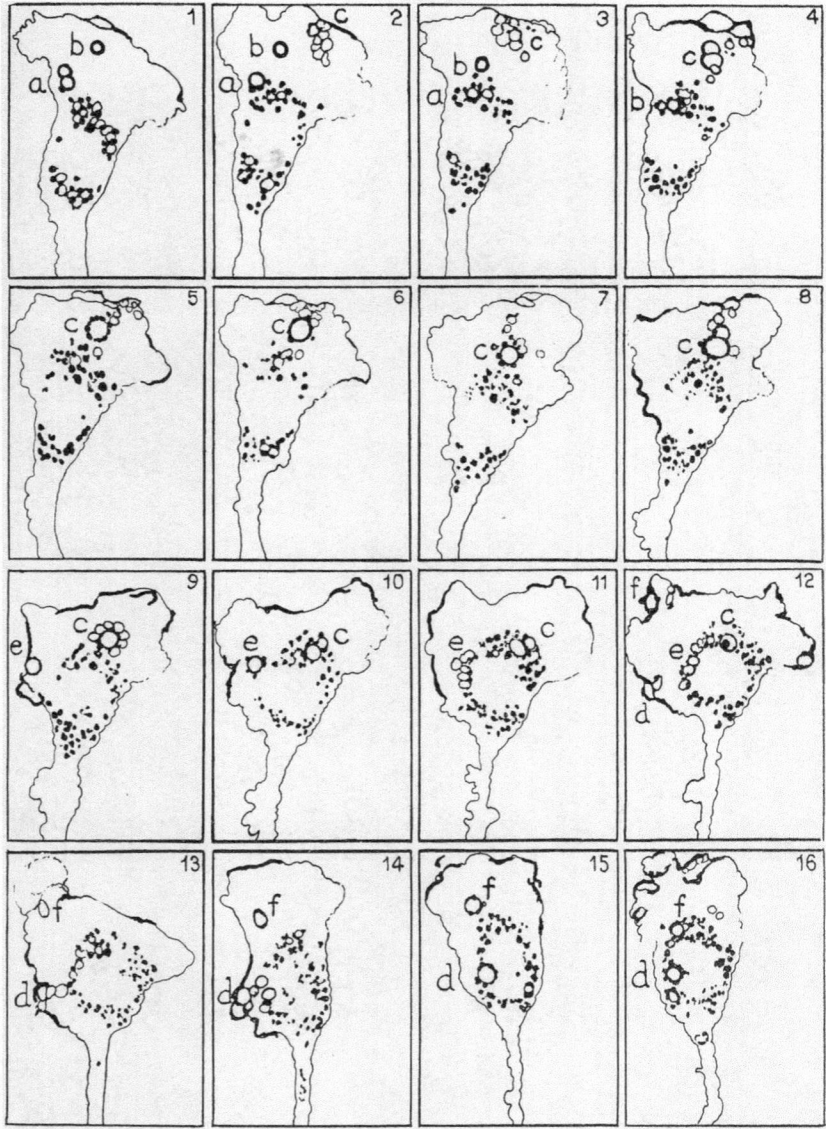

Figure 5.3 Schematic tracing from the film stills of "Pinocytosis" to show the movement of fluid-containing vesicles from the surface of the cell to its interior. (Lewis, Warren H. Pinocytosis. *The Johns Hopkins Hospital Bulletin* 49 (1929), 26, 27. © The Johns Hopkins University Press. Reprinted with permission.)

Changes in form of the individual cells are continuously taking place along with continuous changes of the pseudopodia . . . The periphery is like a ruffle the free edge of which is much longer than the more proximal circumference, and consequently it is thrown into a series of folds . . . The fluid globules which are taken up at the periphery of the wavy veil-like membranous pseudopodia seem to get caught or trapped in the folds of the ruffle, which probably then fuse around and completely enclose them . . . One cannot see exactly how the folds enclose and fuse about a globule of fluid, but there is no doubt about the fact that globules do actually get into the cell and then move rapidly towards the center.[42]

The scale of this activity was also noted. Lewis wrote that the cells sometimes took in a "relatively enormous" volume; sometimes one vacuole after another, and often several at a time, could be seen passing from the periphery to the central part of the cell. While it was not exactly clear what happened to these vacuoles – they could be seen decreasing in size and gradually disappearing – they seemed to move in a very purposeful way toward the center of the cell. He calculated that the total volume of such vacuoles might in the course of an hour amount to one-third the volume of the cell, and in twenty-four hours, several times the volume of the cell itself. Since the cells did not actually increase in size, it had to be assumed that the fluid passed out as well as into the cell.

Furthermore, once Lewis had noted this phenomenon in macrophages on film, he also saw it in cultures of rat sarcoma cells and rat carcinoma cells. "Pinocytosis may be a much more universal process than we at present suspect." Cinematography had in this case, through the exaggeration of scale and speed, highlighted a phenomenon which Lewis then began to see everywhere in his tissue cultures: "Motion pictures of these cells migrating on the coverglass revealed for the first time pinocytosis. Some one would no doubt have ultimately observed it by following cells in the ordinary manner, for, after we had seen it in the motion pictures, we were able to follow it under the microscope."[43] Cinematography revealed not just levels of activity hitherto unsuspected, but also specific new forms of activity, which, even if they were in principle visible through the microscope, were not noticed until exaggerated by microcinematography. Some of these were useful for telling one type of cell from another; others would turn out to be universal features of cellular life, like pinocytosis. This mechanism of interaction between the cell and its surrounding medium revealed an "economy of intercellular fluids," which pointed to a remarkably dynamic interchange between the inside and the outside of the cell, and a very active role for the bounding membrane of the cell itself.

This finding is a good example of the always doubly generative nature of early work in tissue culture. As Ross Harrison observed in 1927, very early on the term "tissue culture," came to denote both a laboratory technique and a field of knowledge generated by that technique.[44] To know more about

living cells was to know more about how to keep them alive better outside of
the body. In the case of pinocytosis, this knowledge was built back into tissue
culture method by a scientist called George Gey, who began collaborating
with the Lewises in 1922, but only came to Johns Hopkins in 1929 as Director
of the Tissue Culture Laboratory in the Department of Surgery. Gey (who
worked closely with his wife Margaret Gey) was particularly interested in
growing human cells. Human cells were very difficult to maintain alive outside of the body for more than a few days. Carrel in 1912 had only succeeded
in keeping human sarcoma cells alive for a few days at the most.

While the connection is not explicitly drawn by Gey and Lewis themselves,
the discovery of phenomena such as pinocytosis drove the further development of tissue culture technology in the distinctive direction of incorporating
movement into the apparatus surrounding the living cell. The sight of cells
constantly taking in the surrounding medium, which was assumed to be
part of cellular nutrition and respiration, suggested that their lives would be
better maintained if they had constant access to fresh medium. Why hold a
dynamic subject in a static body? The innovation of the "roller tube" was a
direct consequence of the increased knowledge of the dynamic, interactive
nature of cellular life. In 1933, Gey published the results of several years of
tinkering in the Lewis laboratory, a description of an apparatus that would
itself provide a dynamic environment for cultured cells to live in. This was
called the "roller tube" technique of cell culture. The cells were grown in test
tubes that were rotated very slowly, at about one hour per turn, in a specially
built gas-tight cylinder into which a carbon dioxide gas mixture was delivered
by a pulsing gas-delivery apparatus.

> It is possible to bring about a sort of washing action on the growing tissue cells
> by revolving the tube slowly at constant speed, thus allowing the supernatant fluid
> to bathe them constantly . . . The revolving action of the tube not only permits
> the supernatant or washing fluid to come into contact with the growing tissue cells
> at definitely controlled intervals, but at the same time allows a definite period of
> exposure to any gaseous mixture desired during the gas phase of the turn, when the
> cells are covered by a minimum amount of fluid.[45]

As Gey explained it, this set-up was designed to remedy shortcomings of
both the hanging-drop and Carrel flask methods in trying to grow malignant
cells. In other methods, cells which came free of the plasma medium due to
the liquefaction that the cells produced in their surroundings were generally
lost upon sub-culturing. In culturing a fragment of malignant tumor, the con-
nective tissue cells in the fragment usually proliferated in the solid medium
of the culture while the more malignant cells came free and were found in the
supernatant produced by liquefaction. While it was easy enough to maintain
the part that proliferates in the solid medium, "the more active tumor cells,
on the contrary, usually liquefy most types of media, come free very soon
after the primary culture is made, and lie dormant or die in the stagnant

liquefaction droplet."[46] Since it is those malignant cells with the ability to liquefy fibrin that those studying cancerous cells *in vitro* were interested in, static modes of culture meant a loss of the very cells that were wanted.

With Gey's method, by contrast, the researcher could select via this difference in behavior the malignant, free cells by collecting them from the liquid fraction and culturing them in a constantly moving roller tube: "no stagnation of the medium occurs, as there is constantly supplied a fluid medium which rapidly dilutes any toxic products." Not only was this a much more effective method of keeping cancer cells alive indefinitely, it was also much less laborious as large amounts of cells could be cultured in each tube. The glass containers were designed not only as vessels to contain living cells, but as transparent bodies for the life going on inside them. The roller tubes either had round sides "which permits only a fair estimation of the condition of the cells on microscopic examination," or hexagonal tubes which supplied "a more suitable flat surface for microscopic observation." They also had pointed ends into which the cells would sediment when the tube was stood on end or centrifuged. In short, the method allowed a better simulation of bodily conditions: "The washing action, in a way, simulates the flow of the body fluids." With this technique, Gey was able to culture human cancer cells much more successfully than any of his colleagues, and it is this method which he would later use to establish the first widely used human cell line, HeLa, in 1951.[47] The collaborative effort between the Lewises and their younger colleagues such as George Gey, which eventually led to the establishment of one of the most widely used cell lines in the history of biomedical research, is a good example of the kind of impact the Lewises had within the nascent tissue culture community in the 1930s.

Allowance for the dynamism of cellular life depicted by film was thus built in to the apparatus keeping the cells alive. The apparatus itself began to move. It is no coincidence that this apparatus was being developed in the same laboratory where the phenomenon of pinocytosis was being filmed and followed. The constancy and volume of ingestion and expulsion of fluids suggested by the discovery of pinocytosis was part of the realization that the bodily functions of nutrition, gas exchange, pH regulation, and excretion could only be simulated properly if the movement and flow inside living organisms was also mimicked. The hanging drop method had provided a simple closed cell-fluid system; Gey's apparatus put this system into motion. At the same time, it further automated the labor of maintaining the life of cells; instead of changing the supernatant, constant change was provided by the slow rolling mechanism, in vessels large enough to carry out "massive tissue culture."

With film, Warren Lewis also turned back to the embryological questions that he had seemingly left behind, and made the first time-lapse studies of the fertilization and development of the mammalian egg. The questions of cellular change over time were common to the microcosm of the tissue culture and the microcosm of the early embryo. Discussed above is the effect of these films on further developments of the tissue culture technique, but it should also be said that the two effects of film distribution, portraying results to fellow scientists, and reaching a wider popular audience, are not easily teased apart. Among its other qualities, scientific cinematography was lauded by many as an excellent medium for teaching and science popularization. Eastman Teaching Films released the 1929 film of the development of a fertilized rabbit ovum. Warren Lewis wrote to Eastman, protesting the "excessive cost" (of $45.00 for 35 mm and $22.50 for 16 mm) at which the company was making this film available. He wrote because he felt the cost would interfere with the accessibility of these films: "our object is to give them as wide a distribution as possible just as we do with reprints as they represent certain phases of our scientific work which cannot very well be demonstrated in any other manner than by seeing the films themselves."[48]

However, the role of these films was never limited to representing "certain phases" of scientific work. Many people who would never read a reprint of a scientific paper saw them. *Development of the Fertilized Rabbit Ovum* was included in the Hall of Science at the 1933 International Exposition "A Century of Progress." Lewis was besieged with requests for his films, such as the following from J. P Schramm of the Botany Department at the University of Pennsylvania: "So impressed were some of the men in our department with your remarkable films shown at the seminar that they have raised the question whether it would be possible to secure a copy sufficient to show the remarkably dynamic character of nuclear and cell division. Such film, of course, would give even elementary students an appreciation of cell activities which it is difficult in any other way to get over so well."[49] Other requests for films, or permission to use parts of films, came from *Encyclopedia Britannica* (for a film on mitosis), Walt Disney (permission to use the film *Pinocytosis* for the Bell Science Series), and Warner Brothers (who offered $200 in 1960 for the use of four scenes from *Dividing Cancer Cells*).

The figure "Thirsty Cells" (Fig. 5.4) shows the delighted reaction of the newspapers to public showings of films such as *Pinocytosis*. The cartoon figure of the cell is an anthropomorphized blob with a face, hands and feet (wearing shoes), drinking. Headlines also trumpeted the films as though they were in themselves scientific breakthroughs (Fig. 5.5). I see these responses as a popular version of the growing realization in the biological sciences of the cell as an active agent, not an inert component or "building block" of tissues and organs and bodies. The allied reaction in less popular and more pedagogical contexts can also be found. This typical example drawn from a general cytology textbook, in a subsection entitled "Living and fixed cells," shows that cinematography was recognized as a unique way of accessing the life of the cell over time, and that this access had noticeably reoriented conceptions of the cell:

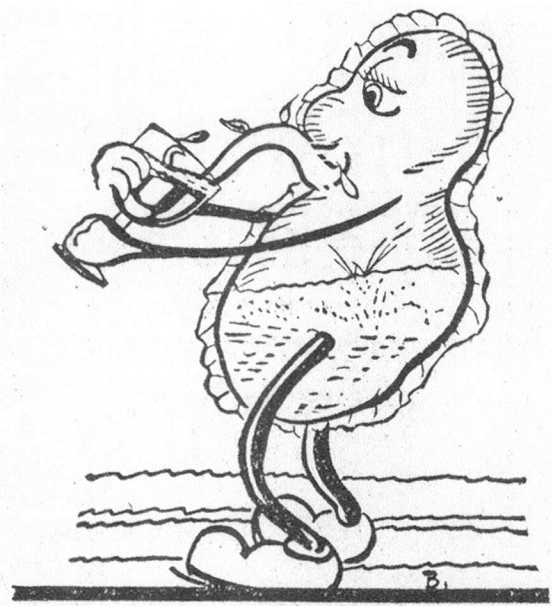

Figure 5.4 "Thirsty Cells." Newspaper cartoon illustrating a 1936 story about the film *Pinocytosis*. (*Baltimore Sun* July 22, 1936.) Artist unknown.

> Of great interest . . . is the recent application of motion-picture photography to cytological problems. Films taken slowly of cells in tissue cultures and then projected at high speeds have served to furnish information, especially with regard to time relations, which it would be difficult or impossible to obtain by any other means. *Moreover, they have placed desirable emphasis on the conception of the cell as a dynamic system.* As a consequence of the renewed study of living cells, it has been possible to evaluate anew the results obtained by the standard methods of fixation and staining and to improve upon such procedures.[50]

It is this emphasis, "on the conception of the cell as a dynamic system," that the Lewises worked towards throughout their whole career. Film both epitomized and communicated this idea, through its own dynamic, circulating, flexible form.

Conclusion

Margaret and Warren Lewis were part of a small community of researchers who dedicated their laboratories and their research careers to developing the technique of tissue culture and the field of knowledge about the dynamic, living cell made possible by that technique. Together with Alexis Carrel at the Rockefeller Institute, the Strangeways Laboratory in Cambridge, England,

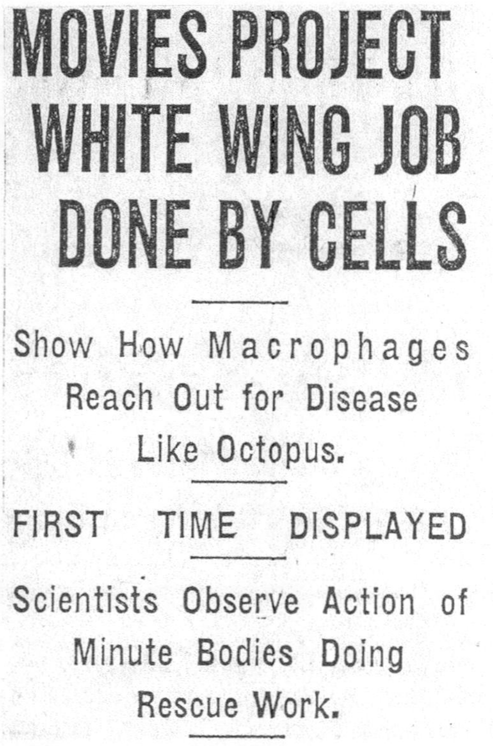

MOVIES PROJECT
WHITE WING JOB
DONE BY CELLS

Show How Macrophages
Reach Out for Disease
Like Octopus.

FIRST TIME DISPLAYED

Scientists Observe Action of
Minute Bodies Doing
Rescue Work.

Figure 5.5   Newspaper headlines after screenings of films of macrophages by Lewis in 1936. Source unknown. (Alan Mason Chesney Archives.)

and George and Margaret Gey at Johns Hopkins, they filled a role of a kind of social and technical backbone of a small field that did not really take off until after World War II, when the availability of antibiotics made growing tissues *in vitro* far easier. The Lewises were thus part of the nucleating force that brought together the professional association called the Tissue Culture Association in 1946.

Not a great deal of historical attention has been lavished on the Lewises to date, but this may be seen as an effect of several different factors. First, historiography of twentieth century biology has been oriented toward genetics and molecular biology, though this may change with the current attention to techniques based on the ability to manipulate cells outside of the body, such as cloning and stem cell technology. Second, the Lewises won no Nobel prizes and made no earthshaking discoveries, and never moved into positions of administrative or political prominence, preferring to remain in the laboratory even after retirement. Warren Lewis was not, by all accounts, a man to push his own accomplishments; Margaret Lewis was not, by virtue of being a woman in early twentieth century science, in a position to push her own accomplishments even if she had been inclined to do so. George Streeter, in writing to Ross Harrison in 1931 to ask for Harrison's support of a nomination of Lewis to the National Academy of Sciences, wrote that "the main handicap of Lewis is that he is not a show man." Streeter's efforts in 1931 met with no success. He tried again in 1933, writing again to Harrison to say "I feel strongly that his life-long devotion to research and the high type of his observations have earned the distinction election would confer. Lewis has never been a very good platform performer and it has resulted in his not being so well known as he should be."[51] It is clear that these individuals were and are highly significant to those who knew them, in particular the many tissue culture practitioners who can trace genealogically – indirectly or directly – the Lewises' impact on their training. There are many, because the Lewises were generous with their time and eager to spread their methodological approach; over such a long career, this adds up to an extended sphere of influence. However, other than this, they are rather obscure figures in the history of twentieth century biology.

Now, it may be expected at this point, after having spent this chapter detailing the Lewis' life and work, that I now make a claim, absolutely familiar to historians, that I have rediscovered the overlooked historical significance of these individuals and their achievements and it is now time for them to receive the recognition which is their due as pivotal figures in the history of science. Even if I do not make this claim explicitly, it might be read as being implicitly present in the choice to lavish such detailed attention on these two and their writings and films. Therefore, I wish to end by proposing a different way to think about the purpose of reading the Lewises' life and work. I actually do not think that they were pivotal figures with immense influence in changing the intellectual direction of science. They followed and further developed what Harrison and Carrel and Burrows first did. They learned to film from others who used time-lapse cinematography to observe tissue cultures and embryos earlier. They entered fields of debate about blood cells and malignancy that were already well occupied. They did excellent and solid work which contributed new things to all of these. To insist on some kind of primacy for their work would be to misunderstand the way in which their work was foundational.

What is most interesting historiographically about the kind of work they did was exactly its tendency to disappear in terms of conventional fame or posterity. This is a story whose detail shows how one person's scientific work can become another's unmarked protocols and assumptions, and it is the nature of assumption not to be held continuously in the foreground. Warren Lewis' observations on his film of pinocytosis sum this up perfectly. He wrote that once he saw the phenomenon exaggerated by film, he began to see it everywhere. Indeed, the whole field of endocytosis, how things get into cells and what happens to them once they are there, could be said to be a result of others beginning to see this phenomenon everywhere. And once something is so apparent everywhere, it is very hard to remember what it was like to not see it. The same is true on a more general scale. This chapter has labored and repeated the idea that it was not natural or given to emphasize the importance of cellular behavior and the importance of finding ways to study it; this merits repetition in part because this assumption has become so seamlessly incorporated into the organization of biological knowledge today. There is something indefinite about results that take the form of a way of seeing. The Lewises contributed much to a shift that they helped begin build, but which was much larger than just their individual efforts – seeing cells as dynamic agents in the life of the body. The Lewises' techniques were not marked as discoveries or breakthroughs, but disappeared due to the fact that they were really useful ways for others to make discoveries and breakthroughs.

Because of this, it is suitable to end on the words of someone who recognized the inestimable quality in the Lewises' work, having learned tissue culture from them herself. Again, on the occasion of Warren Lewis' 85th birthday, tissue culturist Margaret Murray wrote to say

> Congratulations on your many years of fruitful work, most especially in the field of tissue culture. This infant science you fostered from its early beginnings, and with a brilliant simplicity developed its potentialities in manifold directions. All subsequent workers owe to you both an inestimable debt of gratitude. For we have built on the sound foundations laid by 'the Lewises'. This feeling is especially strong in the writer, since it was with Mrs. Lewis' help that I took my first step.[52]

Acknowledgements

I would like to thank Len Schiff, head of the historical committee of the Society for In Vitro Biology for his generous help, Alda Vidrich for talking with me at length about her sense of the general significance of early developments in tissue culture technique, Marnie Halpern for her time and enthusiasm in talking about Warren Harmon Lewis, Patricia Gossel for letting me read her unpublished paper about the Lewises, and Gerald Friedland for sharing his knowledge about Margaret Reed Lewis.

Notes

1. Paul Weiss, "A poem in honor of Dr. Lewis," On the occasion of Warren Lewis' 85th birthday. Dinner program from The Southern Hotel, Baltimore, Maryland, 15 November 1955, CIW Archives, Folder Embryology – "Lewis, Warren."
2. Jane Oppenheimer, "Cells and organizers," *American Zoologist* 10 (1970), pp. 75–88.
3. Gerald Friedland and Bert Thurber, "The pioneering efforts of women in tissue culture during the first 50 years of its development," unpublished manuscript, 1999.

4. George Streeter to John C. Merriam, 29 August 1934, CIW Archives, Folder Embryology – "Annual Reports."

5. Sergey Federoff and Leif Hertz (eds.), *Cell, Tissue and Organ Cultures in Neurobiology* (New York: Academic Press, 1977), p. 1.

6. Sergey Federoff, "Anatomists' contribution to tissue culture," John E. Pauly (ed.), *The American Association of Anatomists, 1888–1987* (Philadelphia, PA: Williams and Wilkins, 1987), pp. 208–19.

7. George Corner, "Warren Harmon Lewis, June 17, 1870 – July 3, 1964," *Biographical Memoirs National Academy of Sciences* 39 (1967), pp. 322–58.

8. Warren H. Lewis, "Experimental studies on the development of the eye in amphibia. I. On the origin of the lens. *Rana palustris*," *American Journal of Anatomy* 3 (1903), pp. 505–36; and Warren H. Lewis, "Lens formation from strange ectoderm in *Rana sylvatica*," *American Journal of Anatomy* 7 (1907), pp. 145–69.

9. Warren H. Lewis, "Transplantation of the lips of the blastopore in *Rana palustris*," *American Journal of Anatomy* 7 (1907), pp. 137–43.

10. Frederick Bang, "History of tissue culture at Johns Hopkins," *Bulletin of the History of Medicine*, 51 (1977), pp. 516–37.

11. Jane Maienschein, *Transforming Traditions in American Biology, 1880–1915* (Johns Hopkins University Press, 1991); Jane Maienschein, "Experimental biology in transition: Harrison's embryology, 1895–1910," *Studies in History of Biology* 6 (1983), pp. 107–27; and Hannah Landecker, "New times for biology: nerve cultures and the advent of cellular life in vitro," *Studies in the History and Philosophy of the Biological and Biomedical Sciences* 33 (2002), pp. 667–94.

12. Ross Harrison, "Observations on the living developing nerve fiber," *Proceedings of the Society for Experimental Biology and Medicine* 4 (1907), pp. 140–3.

13. Alexis Carrel and Montrose Burrows, "Cultivation of tissues in vitro and its technique," *Journal of Experimental Medicine* 13 (1911), pp. 387–96.

14. Pnina G. Abir-Am and Dorinda Outram (eds.), *Uneasy Careers and Intimate Lives: Women in Science, 1789–1979* (New Brunswick, NJ: Rutgers University Press, 1987).

15. Margaret R. Lewis and Warren H. Lewis, "The growth of embryonic chick tissues in artificial media, agar, and bouillon," *Johns Hopkins Hospital Bulletin* 22 (1910), p. 126.

16. Ibid., pp. 126–127.

17. Margaret R. Lewis and Warren H. Lewis, "The cultivation of tissues in salt solutions," *Journal of the American Medical Association* 56 (1910), p. 1,795.

18. Ibid., p. 1795.

19. Margaret R. Lewis and Warren H. Lewis, "The cultivation of tissues from chick embryos in solutions of NaCl, CaCl$_2$, KCL, and NaHCO$_3$," *Anatomical Record* 5 (1910), p. 283.

20. Margaret R. Lewis and Warren H. Lewis, "The contraction of smooth muscle cells in tissue cultures," *American Journal of Physiology* 44 (1917), p. 67.

21. Ibid., p. 68 (emphasis added).

22. Gerald Geison, *Michael Foster and the Cambridge School of Physiology: The Scientific Enterprise in Late Victorian Society* (Princeton, NJ: Princeton University Press, 1978).

23. Henry F. Smyth, "The cultivation of tissue cells in vitro and its practical application," *Journal of the American Medical Association* 62 (1914), p. 1,377.

24. William Seifriz to Albert Ebeling, 20 April 1933, Alexis Carrel Papers, Lauinger Library Special Collections Division, Georgetown University, Correspondence.

25. Margaret R. Lewis and Warren H. Lewis, "Mitochondria (and other cytoplasmic structures) in tissue cultures," *American Journal of Anatomy* 17 (1915), p. 340.

26. Margaret R. Lewis and Warren H. Lewis, "Mitochondria in tissue culture," *Science* 39 (1914), p. 332.

27. Ross Harrison, "The life of tissues outside the organism from the embryological standpoint," *Transactions of the Congress of American Physicians and Surgeons* 9 (1912), p. 64.

28. Margaret R. Lewis, "The destruction of *Bacillus radicicola* by the connective-tissue cells of the chick embryo in vitro," *Johns Hopkins Hospital Bulletin* 34 (1923), p. 224.

29. Ibid., p. 225.

30. Corner, "Warren Harmon Lewis," p. 338.

31. Margaret R. Lewis, "The formation of vacuoles due to Bacillus typhosus in cells of tissue cultures of the intestine of the chick embryo," *Journal of Experimental Medicine* 31 (1920), pp. 293–311.

32. Warren H. Lewis, "The transformation of mononuclear blood cells in macrophages, epithelioid cells, and giant cells," *Harvey Lectures* 21 (1925–6), p. 77.

33. Ibid., p. 78.

34. Corner, "Warren Harmon Lewis," p. 338.

35. Warren H. Lewis, "Malignant cells," *Harvey Lectures* 31 (1936), p. 222.

36. Anonymous, "Hydrophagocytosis," *Journal of the American Medical Association* 95 (1930), p. 1,509.

37. Margaret R. Lewis and Warren H. Lewis, "The contraction," p. 69.

38. Warren H. Lewis, "Malignant cells," p. 222.

39. Ronald Canti, "Cinematograph demonstration of living tissue cells growing in vitro," *Archiv für experimentelle Zellforschung* 6 (1928), p. 87.

40. Warren H. Lewis, "Pinocytosis," *Johns Hopkins Hospital Bulletin* 49 (1929), p. 18.

41. Anonymous, "Hydrophagocytosis," p. 1,509.

42. Warren H. Lewis, "Pinocytosis," pp. 18–19.

43. Ibid., p. 18.

44. Ross Harrison, "The status and significance of tissue culture," *Archiv für Experimentelle Zellforschung* 6 (1927), pp. 4–27.

45. George Gey, "An improved technic for massive tissue culture," *American Journal of Cancer* 17 (1933), p. 752.

46. Ibid., p. 753.

47. Hannah Landecker, "Immortality, in vitro: a history of the HeLa cell line," in Paul Brodwin (ed.), *Biotechnology and Culture: Bodies, Anxieties, Ethics* (Bloomington, IN: Indiana University Press, 2000), pp. 53–72.

48. Warren Lewis to H. Senior, 30 September 1930, Warren Lewis Papers, American Philosophical Society, Correspondence PR–SP.

49. J. P. Schramm to Warren Lewis, 5 March 1941, Warren Lewis Papers, American Philosophical Society, Correspondence PR–SP.

50. Lester W. Sharp, *Introduction to Cytology*, 3rd edn. (New York: McGraw Hill, 1934), p. 7 (emphasis added).

51. Streeter to Harrison, 17 October 1933, CIW Department of Embryology Papers, Alan Mason Chesney Archives, George Streeter correspondence.

52. "On the occasion of Warren Lewis' 85th birthday," p. 4.

Bibliography

Abir-Am, Pnina G. and Dorinda Outram (eds.), *Uneasy Careers and Intimate Lives: Women in Science, 1789–1979* (New Brunswick, NJ: Rutgers University Press, 1987).

Anonymous, "Hydrophagocytosis," *Journal of the American Medical Association* 95 (1930), p. 1,509.

Bang, Frederick, "History of tissue culture at Johns Hopkins," *Bulletin of the History of Medicine* 51 (1977), pp. 516–37.

Canti, Ronald G., "Cinematograph demonstration of living tissue cells growing in vitro," *Archiv für Experimentelle Zellforschung* 6 (1928), p. 86–97.

Carrel, Alexis and Montrose Burrows, "Cultivation of tissues in vitro and its technique," *Journal of Experimental Medicine* 13 (1911), pp. 387–96.

Corner, George W., "Warren Harmon Lewis, June 17, 1870 – July 3, 1964," *Biographical Memoirs National Academy of Sciences* 39 (1967), pp. 322–58.

Federoff, Sergey, "Anatomists' contribution to tissue culture," in John E. Pauly (ed.), *The American Association of Anatomists, 1888–1987* (Philadelphia, PA: Williams and Wilkins, 1987), pp. 208–19.

Federoff, Sergey and Leif Hertz (eds.), *Cell, Tissue and Organ Cultures in Neurobiology* (New York: Academic Press, 1977).

Friedland, Gerald W. and Bert D. Thurber, "The pioneering efforts of women in tissue culture during the first 50 years of its development," unpublished manuscript, 1999.

Geison, Gerald L., *Michael Foster and the Cambridge School of Physiology: The Scientific Enterprise in Late Victorian Society* (Princeton University Press, 1978).

Gey, George, "An improved technic for massive tissue culture," *American Journal of Cancer* 17 (1933), pp. 752–6.

Harrison, Ross, "The life of tissues outside the organism from the embryological standpoint," *Transactions of the Congress of American Physicians and Surgeons* 9 (1912), pp. 63–75.

"The status and significance of tissue culture," *Archiv für Experimentelle Zellforschung* 6 (1927), pp. 4–27.

Observations on the living developing nerve fiber," *Proceedings of the Society for Experimental Biology and Medicine* 4 (1907), pp. 140–3.

Landecker, Hannah, "New times for biology: nerve cultures and the advent of cellular life in vitro," *Studies in History and Philosophy of the Biological and Biomedical Sciences* 33 (2002), pp. 667–94.

"Immortality, in vitro: a history of the HeLa cell line," in Paul Brodwin (ed.), *Biotechnology and Culture: Bodies, Anxieties, Ethics* (Bloomington, IN: Indiana University Press, 2000), pp. 53–72.

Lewis, Margaret R., "The destruction of *Bacillus radicicola* by the connective-tissue cells of the chick embryo in vitro," *Johns Hopkins Hospital Bulletin* 34 (1923), pp. 223–6.

"The formation of vacuoles due to *Bacillus typhosus* in cells of tissue cultures of the intestine of the chick embryo," *Journal of Experimental Medicine* 31 (1920), pp. 293–311.

Lewis, Margaret R. and Warren H. Lewis, "The contraction of smooth muscle cells in tissue cultures," *American Journal of Physiology* 44 (1917), pp. 67–74.

"Mitochondria (and other cytoplasmic structures) in tissue cultures," *American Journal of Anatomy* 17 (1915), pp. 339–401.

"Mitochondria in tissue culture," *Science* 39 (1914), pp. 330–3.

"The growth of embryonic chick tissues in artificial media, agar, and bouillon," *Johns Hopkins Hospital Bulletin* 22 (1910), pp. 126–7.

"The cultivation of tissues in salt solutions," *Journal of the American Medical Association* 56 (1910), pp. 1,795–6.

"The cultivation of tissues from chick embryos in solutions of Na Cl, CaCl₂, KCl, and NaHCO₃," *Anatomical Record* 5 (1910), pp. 277–93.

Lewis, Warren H., "Malignant cells," *Harvey Lectures* 31 (1936), pp. 214–34.

"Pinocytosis," *Johns Hopkins Hospital Bulletin* 49 (1929), pp. 17–27.

"The transformation of mononuclear blood cells in macrophages, epithelioid cells, and giant cells," *Harvey Lectures* 21 (1925–6), pp. 77–112.

"Lens formation from strange ectoderm in *Rana sylavatica*," *American Journal of Anatomy* 7 (1907), pp. 145–69.

"Transplantation of the lips of the blastopore in *Rana palustric*," *American Journal of Anatomy* 7 (1907), pp. 137–43.

"Experimental studies on the development of the eye in amphibia. I. On the origin of the lens. *Rana palustris*," *American Journal of Anatomy* 3 (1903), pp. 505–36.

Maienschein, Jane, *Transforming Traditions in American Biology, 1880–1915* (Johns Hopkins University Press, 1991).

"Experimental biology in transition: Harrison's embryology, 1895–1910," *Studies in History of Biology* 6 (1983), pp. 107–27.

Oppenheimer, Jane, "Cells and organizers," *American Zoologist* 10 (1970), pp. 75–88.

Sharp, Lester W., *Introduction to Cytology*, 3rd edn. (New York: McGraw Hill, 1934).

Smyth, Henry F., "The cultivation of tissue cells in vitro and its practical application," *Journal of the American Medical Association* 62 (1914), pp. 1,377–81.

6

HEREDITY, DEVELOPMENT, AND EVOLUTION AT THE CARNEGIE INSTITUTION OF WASHINGTON

GARLAND E. ALLEN

Department of Biology, Washington University

Introduction

In the first five chapters we have seen how the Carnegie Institution of Washington (CIW) Department of Embryology was organized in 1914 and some of the major characteristics of the research carried out there during its first four decades. We have also seen how, in many ways, this department is representative of the underlying principles of organization established by the CIW (the "Carnegie style", as some have called it), and how that style promoted the innovative work that has characterized the Department throughout its ninety-year existence.

A decade earlier a similar philosophy had prompted the CIW to fund the establishment of the Station for Experimental Evolution (SEE) at Cold Spring Harbor, New York, under the aegis of the young and enthusiastic Charles B. Davenport. (The SEE would later be re-named the Department of Genetics, in 1921, reflecting the major work it had been carrying out from the beginning, namely in Mendelian genetics.) The two departments co-existed with relatively little interaction until the Department of Genetics was formally closed in 1972, each pursuing quite independently its own research agenda. Given that embryonic development and what we today call genetics (i.e., patterns of transmission of hereditary material from parent to offspring) were considered in the late nineteenth century to be a single, unified problem, it is curious that the CIW supported the two as separate fields in separate institutions. The earlier chapters in this volume have described research during the early history of the Department of Embryology, so this chapter will focus on the early history of the Department of Genetics as a means of exploring the relationship between the study of heredity and the study of development within the life sciences as a whole at the turn of the twentieth century. This will set the stage for the final two chapters in this volume, which show the variety of ways embryology and genetics had already begun to re-merge, in part through work carried out at the Department of Embryology during the second half of the twentieth century.

Before examining in detail the development of the Department of Genetics, it will be useful first to review, as background, the major changes occurring within the field of biology between 1880 and 1914. This was a period of particular ferment that had a profound impact on the course of genetics and embryology in the early decades of the twentieth century.

Overview: embryology and genetics, 1880–1930

In a highly influential article of 1978, Scott Gilbert emphasized that many of the early pioneers of modern genetics – Edmund Beecher Wilson, Theodor Boveri and Thomas Hunt Morgan – had begun their professional lives as embryologists (he could also have mentioned William Bateson, Charles B. Davenport, and W. E. Castle, all of whom received their Ph.D.s doing traditional embryological work).[1] At the time, much embryological work was subsumed under the general heading of "morphology," whose main aim was to trace out phylogenies using the descriptive methods of comparative anatomy, physiology, and especially embryology. The chief spokespersons for morphology by the 1890s were the German biologists Ernst Haeckel and August Weismann. A number of scholars have written on the nature of Haeckel's and Weismann's integrated theories, so I will provide only a brief summary here.[2] The elaborate scheme put forth by Haeckel in his *Generelle Morphologie* [*General Morphology*] of 1866 (expanded in publications of 1872 and 1874), and by Weismann in his essay *Das Keimplasm, Eine Theorie der Vererbung* [*The Germ-plasm, A Theory of Heredity*] in 1892, were attempts to bring together the problems of heredity, development, and evolution within a single, comprehensive explanatory framework. The details of these schemes need not concern us here, but both Haeckel and Weismann postulated some material, hereditary particles (plastidules for Haeckel; ids, idants and biophors for Weismann). These particles, through being differentially parceled out, or activated, during embryogenesis, could account for differentiation. Through various mechanisms of competition among the particles (Weismann) or influences from the environment (Haeckel) they could provide new raw variation as the material for evolution. Haeckel also tied his scheme into the historical development of lineages, or phylogeny, through his famous "biogenetic law." That ontogeny (development of the individual) could be seen as recapitulating in telescoped form the phylogeny (historical sequence of adult ancestors) of a whole lineage suggested that detailed embryological studies – including even the earliest stages of cleavage – could

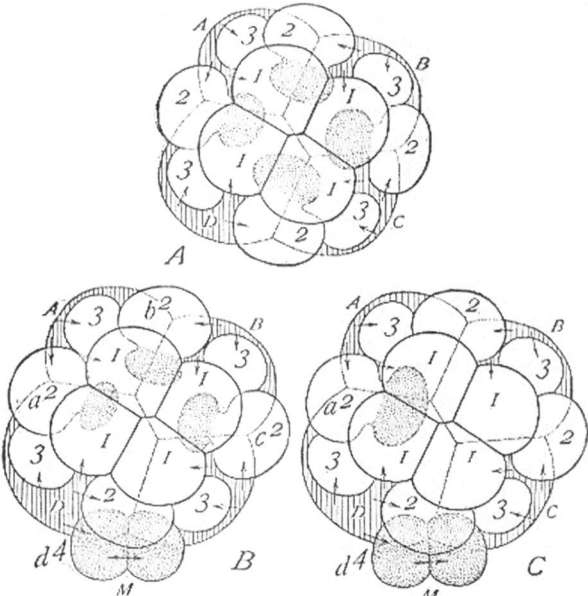

Figure 6.1 Cell lineage in several different mollusks, showing the origin and direction of movement (indicated by arrows) of progeny cells during early embryogenesis (blastula formation). A. *Leptoplana* (a polyclade), B. *Crepidula* (a gastropod), and C. *Unio* (a bivalve) (From E. B. Wilson, *The Cell in Development and Inheritance* (New York, Macmillan, 2nd edn., 1911), Figure 188, p. 414).

provide a means of investigating phylogenetic relationships (Fig. 6.1). Despite their highly speculative nature, morphological theories provided a generation of investigators with a highly suggestive research program. It was not the only type of embryological work being carried out, as we will see later, but it was prevalent enough that many of those who would later become practitioners of the new Mendelian genetics were trained as morphologists, and often reacted strongly against its predominance as a research program.

Gilbert reminds us that issues emerging in what we today would recognize as developmental biology set the stage (in a variety of ways, both positively and negatively) for questions in the new science of genetics that came to the fore with the rediscovery of Mendel's work in 1900. As such, genetics in the early twentieth century had embedded within it – albeit in an unformed state – a set of unresolved embryological problems that were to resurface only later, in the second half of the century. Those problems centered around the question of how genes, once transmitted, exert their effect, and how they do so differentially during morphogenesis: in other words, the physiological functioning of genes on the one hand, and their differential regulation during development on the other. Although raised repeatedly enough during the first four decades of the century, attempts to apply the principles and findings of Mendelian genetics to developmental questions yielded little, either conceptually or methodologically, that could drive a significant experimental research program. Thus, what had been an integrated research problem in the 1890s – that is, the transmission of units of heredity (by whatever name they were called), their selective activation during embryonic differentiation, and their mechanisms of variation – became divided into separate fields: genetics (concerned primarily with mechanisms of variation, transmission, and the material organization of the genes on the chromosomes), and embryology (concerned with tissue- and organ-level changes during development).[3] It was not until the late 1950s, when embryologists began to take Mendelian genetics seriously and had acquired the tools and techniques of molecular biology, that they could begin to see genetics as a paradigm that offered concrete, experimental means of asking developmental questions.

The parallel histories of CIW Departments of Embryology and Genetics thus reflect much broader trends within the biological community as a whole. Those trends include reaction to the synthetic and speculative theories of heredity and development put forward by Haeckel, Weismann, and their contemporaries; a corresponding recognition that embryology offered a variety of interesting questions of its own; the successful introduction into embryology and genetics of experimental methods, which workers in both fields agreed provided greater rigor to their conclusions; and, finally, institutional and professional competition that tended to divide the two fields into separate communities of investigators, often forcing them into turf battles for such resources as research funds, graduate students, and institutional support.

So prevalent was the morphological program in late nineteenth century biology that two future CIW investigators, Thomas Hunt Morgan and Charles B. Davenport,[4] wrote Ph.D. dissertations on morphological topics (Figs. 6.2 and 6.3). Both theses involved detailed microscopic descriptions of changing cell and tissue layer relationships in development. Morgan's investigation was more aimed toward resolving a specific phylogenetic problem (whether the Pycnogonids, or sea spiders, were more closely related to crustaceans or to arachnids) than Davenport's, but both bore the distinct mark of morphological aims and methods: a high level of descriptive detail, considerable speculation about possible significance of the described results, and the overarching importance of evolutionary relationships around which the studies were organized.

The objections raised to these large-scale theories have been discussed by various authors and involve a number of specific points.[5] The most general criticism was that because of their breadth and the apparently arbitrary nature of many of their assumptions and postulated mechanisms, morphological schemes like Haeckel's and Weismann's simply could never be tested: any one

Figure 6.2 Thomas Hunt Morgan, *c.* 1894–5, Berlin. (Courtesy of the late Isabel Morgan Mountain.)

of several phylogenies could invoke the same observations, with no available means of distinguishing between them. Moreover, important questions about the mechanics of embryonic development or more detailed studies of the stages of embryogenesis, especially in higher animals, were being ignored as the embryo was viewed largely as a tool for phylogenetic reconstruction.[6]

Two research traditions within what we would call today developmental biology co-existed in the late nineteenth century, each with a focus on the embryo as the proper object of study: medical embryology and experimental embryology.

Medical embryology, going back through the work of Karl Ernst von Baer in the early nineteenth century to that of Marcello Malphigi and William Harvey in the seventeenth century, sought to understand the intricate changes in the mammalian embryo during the course of development. For practical, as well as religious, ethical, and legal reasons, most of the early stages in human embryogenesis remained uncharted territory,

Figure 6.3   Charles B. Davenport, Director of the CIW Station for Experimental Evolution/Department of Genetics at Cold Spring Harbor, about 1910. (Courtesy, Eugenics Image Archives, Cold Spring Harbor Laboratory.)

and consequently much medical advice to pregnant women had long been based on rudimentary anatomical and physiological information, and on folklore. It was explicitly for the purpose of organizing a research tool to fill in such gaps that Franklin Mall obtained Wilhelm His' embryo collection and added to it when he sought to establish the CIW Department of Embryology in 1914. Thus, medical embryology was a largely descriptive area of investigation (with much comparison among various mammalian groups) aimed at constructing a detailed understanding of the sequence of stages in human embryogenesis that could be used in both a clinical and a laboratory setting.

A second research tradition in late nineteenth-century embryology was more experimentally oriented, and is associated with the names of Wilhelm Roux, Hans Driesch and the "Entwicklungsmechanik" program of the 1880s and 1890s. This approach was particularly attractive to a younger generation of biologists (born after 1865) who were interested in making biology more experimental and rigorous, like physics and chemistry. To them this meant introducing a focus on proximal (how) rather than ultimate (why, or historical) causation, the formulation of testable hypotheses, the use of controlled experiments to distinguish between alternative hypotheses, the collection of quantitative (rather than merely qualitative) data, and an emphasis on physico-chemical and mechanistic explanations. Although these features had long been a part of the methodology of physiology, they had been employed much less frequently in areas such as natural history, embryology, anatomy, evolution, and heredity.

It was precisely in embryology and heredity that these new trends initially found their most successful and exuberant application. The new "experimental embryology" of Roux and Driesch, and the newly rediscovered work of Gregor Mendel on heredity (called "genetics" by William Bateson in 1906) focused on specific hypotheses that could be tested experimentally. For example, the embryologist could ask: "Is embryonic differentiation the result of unequal parceling-out of hereditary material during successive cleavages of the egg?" This hypothesis could be tested, as Roux did, by killing one of the first two blastomeres of the frog egg and observing the effect on the resulting embryo: if the hereditary particles were distributed unequally, only a half, or partial embryo should result (Roux *did* get half-embryos but for different reasons than he thought at the time); alternatively, if both blastomeres receive the same complement of particles, two complete embryos should result (in a counter-experiment, Driesch got two embryos from *separating* the blastomeres of a sea urchin embryo, though also for different reasons than were recognized at the time). Similarly, the geneticist could formulate hypotheses about the genotype of an individual and test them by backcrosses to homozygous recessives, a homozygous parent producing all dominant offspring, and a heterozygous parent producing one-half dominant and one-half recessive offspring. Using such simple, but rigorous methods of framing and testing hypotheses, a generation of "young Turks" set out to make biology into a hard experimental science, standing on what some saw as the same footing as chemistry and physics.[7]

For both T. H. Morgan and C. B. Davenport, the rediscovery of Mendel's work provided just such a new opportunity. Morgan's multitude of earlier experimental studies on development in frogs, earthworms, sea urchins, and the like had come to relatively little in the way of important new principles. Likewise, Davenport's early morphological work had been solid, but undistinguished. Both became avid Mendelians in the first decade of the century: Davenport by 1908, Morgan by the time of his first *Drosophila* paper in 1910.[8] For both men the new field offered the opportunity to study a problem *experimentally*.

Mendelian genetics also offered a form of analytical method that seemed to Davenport, Morgan, and their contemporaries more akin to the approach taken in the physical sciences. Complex systems could be analyzed into their simpler components, each of which could then be studied under more controlled conditions. For example, the Mendelian geneticist could think of germ cells as carriers of discrete hereditary factors (genes, with their various allelic forms) that were combined and recombined in a variety of predictable ways by the experimental breeder. This resolution of the organism into a mosaic of discrete parts fitted well the mood of mechanistic materialism so prevalent in the sciences around the turn of the century. It took its cue from the success of the atomic theory in physics and (especially at the time) chemistry. (Debates about the quantum view of the atom were largely lost on biologists in the early decades of the twentieth century.) Indeed, a number of early Mendelians, including William Bateson, W. E. Castle, E. M. East, and Wilhelm Johannsen, repeatedly emphasized the similarity between Mendelian genes and atoms in the specific numerical ways in which they combined, and how they emerged from combinations (through segregation during gametogenesis) unaltered by association with other genes in the parental germ cells.[9]

It is in this context – the enthusiasm to apply experimental methods to all sorts of new biological problems, and specifically the availability of a potentially powerful new paradigm of heredity (Mendelism) – that Davenport approached the CIW in 1902 to fund his own vision of a research station, the Station for Experimental Evolution (SEE), specifically aimed at studying heredity and its application to experimental (selective) breeding. With his morphological background, driving energy, and enthusiasm for the new genetics, Davenport seemed to be the perfect person to head up a research facility at the cutting edge of a new discipline. He agreed wholeheartedly with the CIW principles as outlined by the Executive Committee: cooperation in research supported by a strong organizational base. Davenport was, indeed, enunciating not only the CIW philosophy, but also that of the growing professional community of biologists who were at the time organizing themselves into professional societies and institutions, and starting journals devoted to various specialties.[10]

Davenport's vision for experimental evolution

In the minutes of incorporation of the CIW, one of the major topics of discussion was the extent to which funds should be allocated to individuals working in their own existing institutions, and to what extent funds should be used to establish institutions run, or at least controlled, by the CIW.[11] The decision was made to establish both, with a series of specially created institutions directed by first-rate scientists centered around particular problems, and with grants to exceptional individuals working in their own institutions. The fledgling field of heredity profited from both in the form of Davenport's Station at Cold Spring Harbor, and T. H. Morgan's *Drosophila* laboratory at Columbia (and in grants to other individuals along the way as well).

Charles B. Davenport was by all measures an exceptional person. His Ph.D. in descriptive morphology had led him to an instructorship at Harvard (1893–9), followed by part of a sabbatical year in London with Karl Pearson and Francis Galton learning the new methods of "biometrics" (the application of measurement techniques and statistical analysis to the understanding of the inheritance of traits – for example, height between parents and offspring). Although Davenport did not employ a lot of biometrical analysis in his own work, he was sufficiently versed, and enthusiastic about, the approach to have served for several years on the editorial board of Galton and Pearson's journal *Biometrika*, and to have authored a manual, *Statistical Methods in Biology, Medicine and Psychology*, which was first published in 1899 and went through four editions by 1936.[12] A biologist through and through (unlike Pearson, who remained primarily a statistician), Davenport appears to have viewed biometrics primarily as a means to an end – that is, in investigating matters of variation and heredity – rather than as a matter of interest in its own right.[13]

Davenport also explicitly promoted the idea of experimentation in biology, teaching a course at Harvard in the late 1890s on experimental morphology that resulted in his first book by the same title.[14] This two-volume work covered a wide variety of topics based largely on work he carried out in the laboratory with his students: acclimatization of animals to high temperatures (with W. E. Castle), and to poisons (with H. V. Neal), heliotaxis (with Walter B. Cannon), and geotaxis (with H. Perkins), along with a variety of studies on the effects of chemicals on activity of protoplasm and on growth. From 1905, he was also involved with the American Breeders Association (ABA), a group of academic biologists interested in applying general principles of heredity and evolution who sought to work with practical breeders interested in scientific agriculture.[15] In the meetings and subsequent publication from the ABA, Mendelism figured early on, and prominently.[16] Unlike his British mentors Galton and Pearson, Davenport saw no contradiction between biometry and Mendelism, and remained an advocate of both methodologies throughout his career.[17]

In 1902, when the CIW was first incorporated, Davenport immediately applied for funding. He had already been at the University of Chicago for two years, where he had been promised a farm site for experimentation on heredity in animals. When it appeared that plan was not going to materialize in the immediate future, Davenport began to look elsewhere for funding, and the CIW appeared a most likely possibility.

Davenport's plan, as presented to the CIW in various forms between 1902 and 1904, remained centered on the need for an institute where breeding, hybridizing, and selection experiments could be carried out on a wide variety of organisms, including chickens, rats, goats, land snails, spiders, guinea pigs, and lady-bird beetles.[18] In principle, the invertebrates could be grown and cultured in the laboratory, but the larger animals needed more space and special housing, a point Davenport used to justify the need for a station with available land. His proposal of the site of Cold Spring Harbor, Long Island, was dictated by his long-standing association with the Biological Laboratory of the Brooklyn Institute of Arts and Sciences, where he had taught in the summers and which he later took over as Director (while simultaneously directing operations at the CIW Department).

The main rationale for his focus on breeding and selection was to place the study of evolution on a solid, experimental, and quantitative footing. As he wrote to Henry Fairfield Osborn about his proposal in 1902: "I think the work [at the SEE] ought to be limited pretty closely to experimentation and a careful, usually quantitative, study of results. There are a number of subjects of great importance for evolution, such as the statistical study of variation and geographic distribution which must be distinctly avoided in order that energy shall not be directed away from the main point."[19] What he appears to have meant by this last remark is that the statistical (as opposed to experimental) study of topics such as variation or geographic distribution, as important as they may be for answering certain evolutionary questions, would not become a part of the SEE's work. The focus was to remain single-mindedly on *experimental* aspects of evolution, such as the effects of selection, hybridization, and specific environmental factors on variation in a population, and how the distribution of that variation can be changed over time by preferentially selecting certain phenotypes. Because of his wide-ranging interests and his background, he saw himself as the ideal person to direct such an institution. He wrote to John Shaw Billings, Director of the New York Public Library, and a Trustee of the new CIW, outlining his suitability for the job, in 1902: "I am the author of the only book devoted to the statistical study of Evolution . . . and on Experimental Morphology. I believe I am merely stating the cold truth when I say that in training and in work accomplished in the study of heredity and variation – the elements of evolution – no one in the country is as well prepared for experimental and quantitative studies in evolution as I . . . I am about to spend four months in Europe investigating all Experimental Evolution Stations there . . . to better fit myself for the work of directing the Station for Experiments on Evolution [*sic*], whenever the Carnegie Institution establishes it."[20] It is clear Davenport saw himself as transforming old style morphological concerns into the modern experimental and quantitative – we might say *analytical* – mode of thought. He was making a transition from evolution seen as outcome

Figure 6.4   The main administration building (upper right of center) and the animal care facilities (lower left of center) at the Station for Experimental Evolution, Cold Spring Harbor, about 1920.

(phylogeny, geographic distribution) to evolution as process (transmission and selection).

Establishment and early years of the SEE

Davenport's initial application in 1902 was not funded, for reasons that are not completely clear, but that more than likely grew out of the CIW's pending offer to take over the then financially ailing Marine Biological Laboratory in Woods Hole, a plan that would have scuttled the need for another seaside laboratory at Cold Spring Harbor. (It seems unlikely that the CIW lacked interest in the problem of heredity and selective breeding, since at the same time they were considering granting, and ultimately did grant, Luther Burbank, the "plant wizard," of Santa Rosa, California, $ 10,000 per year for his work on hybridization, grafting, and selecting for new varieties of commercial fruits and vegetables.)[21] Davenport quickly submitted a revised proposal, which was approved; on December 12, 1903, he was awarded the sum of "$34,250, with fixed annual appropriations "to continue indefinitely, or for a long time."[22] In January, 1904, he set about hiring staff and plan facilities at the Cold Spring Harbor site (Fig. 6.4). Despite all his talk about keeping a clear focus, however, Davenport appears to have had little real idea what the Institute's long-term goal would be. In March, he had written to the CIW Executive Committee that: "I have little notion of just what we shall do. We shall reconnoiter the first year."

As was his wont, Davenport immediately devoted considerable amounts of time to micromanaging, focusing on small details such as traveling to New York City to purchase canaries, finches, long-tailed fowl, and a Manx cat (for what reason was not made clear). On the hiring front, he acquired the services of two full-time research assistants; Frank Lutz to "make a reconnaissance of the variability offered by the animals of Long Island" and George H. Shull (an early student of Mendelian hybridization in corn) to "reconnoiter the field of plant variability and gather seed plants for heredity experiments . . ."[23] He also appointed a number of associates who would carry out research projects primarily at their home institutions, but often worked in the summers at Cold Spring Harbor. These included his former mentor E. L. Mark, his former student W. E. Castle (Harvard), and colleagues Raymond Pearl (Johns Hopkins), and E. B. Wilson (Columbia). By contrast to the Department of Embryology, founded just a few years later with a much clearer set of goals, it is surprising that the CIW was willing to fund Davenport to such an extent with so little specific plans laid out.

While it is not the main purpose of this paper to trace the history of the SEE in detail, it is important to note that initially Davenport and the CIW had high hopes for its success as a prototype of the independent, full-time research institution (not unlike German "Institutes" or the Rockefeller Institute for Medical Research). In the first few years, Davenport made some promising appointments: A. F. Blakeslee (cytogenetics of *Datura*, the jimsonweed, and later *Oenothera*, the evening primrose), Oscar Riddle (an early twentieth-century American endocrinologist), George Harrison Shull (an early student of hybridization in corn), Charles W. Metz (a pioneer in the genetic analysis of chromosome groups in species of *Drosophila* other than *melanogaster* – the standard lab model used by the Morgan group and others), and E. Carlton MacDowell (who carried out a long-term study of the inheritance of alcoholism in rats and its possible Lamarckian transmission).[24] However, the early promise was not fully realized under Davenport's directorship, as a series of problems arose over the years that significantly undermined the Department's progress and reputation. When Davenport retired in 1934, the scientific work had ebbed to its lowest level. But the early years held out quite different promise.

The era of the gene: genetics and eugenics at Cold Spring Harbor

Mirroring changes within the larger biological community as a whole, by 1908 Davenport had made the intellectual break from experimental morphology to a growing interest in Mendelian genetics and its application to the breeding of mice and, especially, poultry. He authored a series of monographs on the genetics of coloration and plumage in chickens that was a landmark in the early years of Mendelism applied to animals. At about the same time, however, he also became increasingly interested in human heredity and eugenics. This changing focus from animals to humans started early on. His first paper on human heredity, co-authored with his wife Gertrude, was concerned with the inheritance of eye-color in humans.[25] Other papers on inheritance of hair form, various physical characteristics such as brachydactyly, Huntington's chorea, and skin pigmentation followed. The study of the inheritance of skin color led to issues of racial crossing, a matter of considerable concern to Davenport and other eugenicists at the time.

By 1910, Davenport's chief interests lay more and more with eugenical issues. Acting on these interests, in that year he convinced Mrs. Mary Harriman, widow of the recently deceased founder of the Union Pacific Railroad, E. H. Harriman, to donate funds to found a second institution at Cold Spring Harbor, the Eugenics Record Office (ERO), on property adjacent to, and up the hill from, the SEE. Since Davenport's work in eugenics has been described elsewhere it need not be discussed in detail here.[26] Its relevance to the topic at hand, however, lies in the transition it marks not only for Davenport himself, but for the institutions he commanded, from the more integrated approach of morphology to the fully atomized approach of Mendelian genetics that separated the problem of heredity from that of embryology in the biological community at large. From 1912 or 1914 onward the focus of all the work at the ERO, as well as increasingly that at the SEE itself, was on transmission genetics, rather than the original and broader area of "experimental evolution." In the concluding section of this chapter I explore in more detail what this separation came to mean for biology in general, and for the biological programs sponsored by the CIW in particular.

As genetics research rapidly developed around the world in the 1920s, Davenport worked to consolidate his administrative operations at Cold Spring Harbor. In 1916, Mrs. Harriman provided an additional endowment for the ERO and it was officially transferred to the CIW.[27] By 1921, both institutions had been combined to form the Department of Genetics, putting it on equal footing with other CIW departments, including Embryology. Administratively, creating a department gave Davenport more direct control, but its most important advantage was most likely that it gave the ERO greater legitimacy in the scientific world. It was no longer so easy to claim, as it had been previously, that the work done there in eugenics was an amateur operation funded by special, monied interests. (In a parallel attempt to legitimize the work of the ERO at the time, its Superintendent, Harry H. Laughlin, who had only a master's degree when he came to Cold Spring Harbor, was granted an Sc.D. degree from Princeton in 1916, under the sponsorship of E. G. Conklin.) While some workers at the SEE continued to investigate physiological aspects of genetics and evolution (for example, Oscar Riddle, who worked on the endocrinology of sex determination and related topics), work at the new Department, as at the ERO, increasingly focused on gene transmission and cytogenetics. At the ERO this was particularly apparent in the family pedigree charts that Davenport and Laughlin constructed for an ever widening array of human traits, including criminality, manic depression, epilepsy, "inherited scholarship," nomadism, pellagra, and feeblemindedness. Virtually all of these conditions were interpreted in Mendelian terms (as dominant, recessive, sex-linked, etc.), and claims were made that they could be eliminated from the population in a few generations by selective breeding. Although controversial, these claims received support from CIW President John C. Merriam. As late as 1928, Merriam wrote in a memo that eugenics fit well the CIW's charter to carry out work "for the improvement of mankind."[28] Merriam went on to elaborate: "The subject of eugenics, if treated broadly, represents one of the most significant of all fields of scientific endeavor . . . It has been the privilege of the Carnegie Institution to enter the field of eugenics in close relation to studies of genetics."[29]

Despite his enthusiasm, Merriam knew that there were problems developing at the Department of Genetics by the early 1920s. Many of these problems had to do with Davenport's idiosyncratic methods of administration, his problematic relationship with Merriam himself and other CIW leadership, his increasing interest in and attention to eugenics over other work at the laboratory, and his unwillingness to work more collaboratively with other CIW departments. In particular, two concerns increasingly seemed to occupy Merriam's attention. One was the overall quality of work being carried out at Cold Spring Harbor. The other, and to some extent related, issue stemmed from his desire to see more collaboration between the Departments of Genetics and Embryology.

By the early 1920s, Merriam was hearing doubts expressed by a number of biologists about the general quality of *all* the work at Cold Spring Harbor. In a memo of October 12–13, 1923, Merriam reported on a discussion he had with T. H. Morgan (who since 1915 had been receiving his own CIW funding at Columbia for *Drosophila* genetics) about the work going on under Davenport's direction:

> Dr. Morgan considers work of Department of Genetics done in considerable measure by inferior men with relatively little coordination and no very significant result. Work of [A.M.] Banta relatively unimportant; [Charles W.] Metz of second rank; Riddle of second or third rank; [J. Arthur] Harris of second or third rank. Work of [A.F.] Blakeslee [on cytogenetics of *Datura*, jimson weed] good; [John] Belling [on cytogenetics of *Oenothera*, the evening primrose] important . . .[30]

Figure 6.5 John C. Merriam (left), President of the CIW and Charles B. Davenport, Director of the Department of Genetics, at Cold Spring Harbor about 1930.

Merriam's notes of discussions at a National Academy meeting in November 1923, with University of Pennsylvania cytogeneticist C. E. McClung (November 1, 1923) and Johns Hopkins geneticist/biostatistician Raymond Pearl (November 10–14, 1923) all seemed to confirm Morgan's assessment. A major issue underlying all these concerns was that the program was too diffuse, involved little interaction among the various investigators, was not focused on any central problem, and was not necessarily carried out by the best people. At that same meeting, J. McKeen Cattell, former Columbia University psychologist and at the time editor of *Science*, voiced similar views even about the eugenics work coming out of Cold Spring Harbor. In no quarters did the professional reputation of the Department appear to be strong.

Merriam discussed these concerns with Davenport on October 19 at Cold Spring Harbor (Fig. 6.5), and in mid-November at the Academy meeting. He tried to convey to Davenport the importance of developing an overall research focus for the Department, while still keeping an eye on the larger issues. It would also be important, he claimed, for other members of the staff to help strengthen the work of the ERO, many of whose investigators were considered below par.[31] There is no indication as to how Davenport responded, but it is apparent that nothing had improved from the fact that a year later, as the Department of Genetics was heading toward its twenty-fifth anniversary, Merriam again echoed the same concerns, including new worries arising from Davenport's increasing involvement with the Long Island Biological Laboratory in Cold Spring Harbor (of which he had also assumed directorship in the mid-1920s).[32] Indeed, in six years Davenport would retire, followed by Merriam, who would be replaced by Vannevar Bush in 1938. Bush had no sympathy with the ERO or with much that was being done in the other research areas at Cold Spring Harbor. In 1938, he set in motion the machinery to force Laughlin's retirement and the ERO itself was closed on December 31, 1939.

With regard to cooperation between the Departments of Genetics and Embryology, on several occasions Merriam raised the possibility with Davenport. In 1923, a memo suggests that Davenport might extend the cytogenetic work of Blakeslee and Belling to include the interaction of chromosomes with "external influences, physical, chemical and biological, upon the cell . . ." He continued, "studies in the control of development or ontogeny might probably be associated with such studies . . ."[33] In 1929, Merriam explicitly urged more collaboration between E. Carlton MacDowell in the Department of Genetics and the Lewises at the Department of Embryology.[34] Again, as far as the record shows, nothing came of these suggestions. Although there were apparently a few joint meetings, starting in 1934, of the newly organized Division of Animal Biology, as Jane Maienschein, in chapter 1, points out, these produced no noticeable cooperative research projects. By contrast, the Department of Embryology developed close working relationships with several departments at Johns Hopkins University Medical School, bringing about, as we have seen (see earlier chapters in this volume) some very fruitful collaborations. Mall's human embryo collection was useful to investigators in a variety of clinical and research settings and it formed the core around which the initial work was organized. Most important, the new work that was added (such as the primate colony) grew naturally out of the research that was initially based on the embryo collection. While there was not a grand plan set up at the beginning, the work at the Department of Embryology had a logic and leadership that contributed to its continual growth and development.

The SEE never had such a focus, something the CIW administrators apparently did not see as a problem in 1904. They invested in Davenport, certainly a creative and energetic investigator, and in an area (experimental breeding and selection) that loomed as one of the cutting edges of research in the early 1900s. But Davenport's personal and intellectual idiosyncrasies ultimately made him ill-suited to direct a major research institution. Not only did he lack an overall and consistent focus, his unwillingness to engage in or build working relationships outside his own set of interests greatly reduced the possibility of producing an integrated research program. It remained effectively a collection of individual workers who interacted very little. Davenport's increasing attention after 1914 to eugenics, with the resulting disregard of his work in experimental genetics, not only marginalized his own research in the eyes of many of his contemporaries, but also drew him increasingly away from the work of the rest of the staff of the Department. Adding to all these problems, he possessed a rigidity in response to criticism that made it difficult to discuss his work or management style productively. Davenport's peculiar inability to assess the quality of work carried out either by himself, or others at the Department, contributed to the overall decline in work at Cold Spring Harbor under his guidance.

Prospects for the future of the Department began to brighten after Davenport's retirement in 1934. Blakeslee filled in as a temporary director through 1941, when activity began to pick up considerable momentum under the directorship of Milislav Demerec (Fig. 6.6), a student of R. A. Emerson at Cornell. Demerec had come to CSH in the mid-1930s as an investigator, having switched from maize to *Drosophila* as his experimental organism. Demerec soon hired another Emerson student, Barbara McClintock (Fig. 6.7), whose work on gene transfer in maize proved to be, by the late twentieth century, among the best-known ever to emerge from the laboratory (McClintock won the Nobel Prize in 1983 for her work on transposable elements in maize).[35] Under Demerec's leadership the laboratory began to promote more systematically its summer meetings, most notably what became known as the Symposia on Quantitative Biology (and their resulting published volumes), which were to have a major impact on fields as diverse as phage genetics and evolutionary biology.

Adding lustre to the Department's growth, in 1950 Alfred D. Hershey (Fig. 6.8) was induced to leave the Chairmanship of the Microbiology Department at Washington University School of Medicine to take up a research post in the Department. Hershey, who was to share a Nobel Prize in 1969 with Salvador Luria and Max Delbrück, became Head of the New Genetics Research Unit when the department was reconfigured in 1962, a post he held until 1972 when the Unit was discontinued (Hershey and McClintock continued to be funded individually at Cold Spring Harbor by the CIW until their retirement). Under Demerec's and Hershey's guidance Cold Spring Harbor became the spiritual, if not actual, home of the new phage genetics. Thus, over the long haul, a great deal of distinguished work emerged from the Department of Genetics in the post-war years. The details of the CIW's decision ultimately to discontinue the Department of Genetics needs to be studied in more detail by historians of science, as it may reveal in more

Figure 6.6   Milislav Demerec, Third Director (1941–60) of the Department of Genetics at Cold Spring Harbor, taken at the Third International Congress of Genetics, Ithaca, New York, 1932. (From the collection of the late Tove Mohr, Norway.)

depth how the relationship between classical genetics, molecular genetics, and development was perceived at the time by both investigators in the fields and CIW administrators. However, it does seem likely that the CIW administration recognized that work in the new areas of molecular biology and genetics, which had been fostered at Cold Spring Harbor in the 1940s and 1950s, were now being pursued in the Department of Embryology and in individual research programs on the control of gene expression during development.

While the Department of Genetics ultimately redeemed itself as a leader in the early years of molecular genetics, what did not emerge was what CIW administrators hoped might be collaborative efforts between the research at

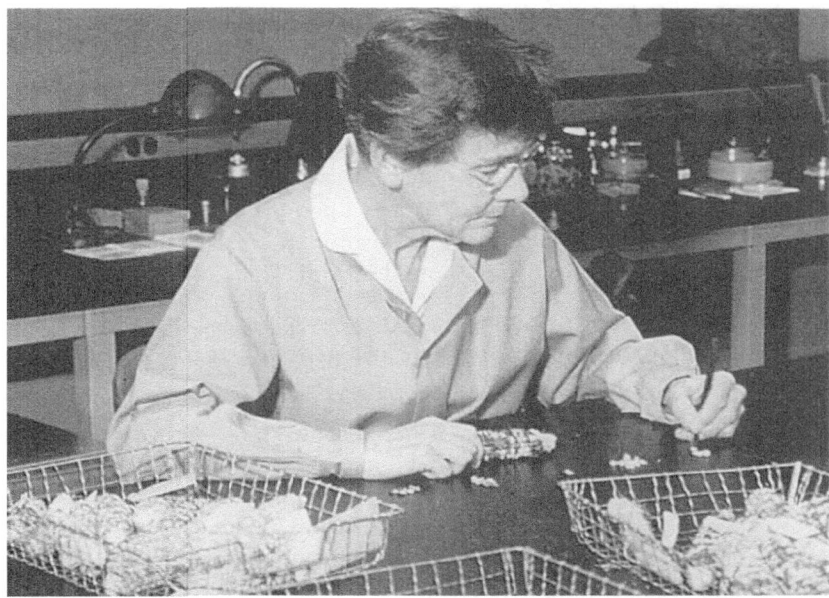

Figure 6.7 Barbara McClintock in her laboratory at CIW Genetics Research Unit, Cold Spring Harbor, *c.* 1963.

Figure 6.8 Alfred D. Hershey in his laboratory at the CIW Department of Genetics, Cold Spring Harbor, about 1960.

Cold Spring Harbor and the Department of Embryology. We now come back to the question about what this lack of interaction signifies beyond local idiosyncrasies associated with Davenport and his management style. What does it tell us about the CIW in particular and the larger biological research community at the time? It might be argued that historical analysis of something that didn't happen is not very interesting. On the contrary, the development of the Department of Genetics permits us to take a closer look at the CIW research style and how an institution seeks to determine what will and will not be part of its supported mission.

Conclusion: embryology and genetics, 1910–50

As we look back at the overlapping histories of the CIW Departments of Genetics and Embryology, several issues become clear. With respect to the "Carnegie style" of supporting research, both reflect the underlying assumption that it is best to fund highly creative individuals rather than pre-designed "programs." Both Davenport in his early years and Mall throughout his career had a commitment to research on a set of problems (Mendelian genetics and human embryology, respectively) that were innovative and "cutting edge" for their day. While the CIW did fund individuals working within their own institutions, such as universities, the decision to set up the SEE or the Department of Embryology showed the enormous potential the Trustees saw in Davenport's and Mall's work at the time. In both cases, the CIW style was to engage the best individuals and turn them loose to build their own institutional base. The strength of this approach is evident in the long history of the Department of Embryology, while the weakness is apparent in the history of the Department of Genetics during the last half of Davenport's reign (roughly 1920–34). The CIW decision eventually to close down the Department of Genetics *per se* in 1962 reflects not any loss of interest in genetics but the realization that, with the rise of molecular genetics (particularly with the work that the CIW had supported at Cold Spring Harbor and elsewhere in the 1940s and 1950s) the field was moving in very different directions. Funding groups working in their individual home institutions became more practical than trying to bring a variety of competitive individuals or groups together under one roof.

More to the point, the history of the two CIW Departments reflects, though was not directly caused by, the growing separation in the twentieth century between the study of heredity as a process of transmission from parent to offspring (i.e., genetics), and heredity as a process of cellular growth and differentiation (embryology). It should be pointed out that as we have seen in the earlier part of this volume, the organization of the Department of Embryology in 1914 grew out of a tradition of medical embryology, associated with Mall's embryo collection, that was very different from the experimental tradition that motivated the establishment of Davenport's Station. What was clear, however, to all parties in the early twentieth century was that the unified study of heredity as *both* transmission and development, as it had been practiced in the 1880s and 1890s, was no longer a viable enterprise. As Mendelian genetics gained a foothold as a major paradigm for investigating patterns of transmission, developmental issues were increasingly set aside, to be pursued by different communities of investigators. To be sure, a few individuals did keep the torch alive for an integrated "developmental genetics:" Richard Goldschmidt, C. H. Waddington, and Ivan Schmalhausen, among others. By the 1920s, however, individual workers or whole research groups were generally identified as *either* geneticists *or* embryologists. The existence of the two Carnegie Departments, with such different and largely non-overlapping research programs, is an indication of just how different the fields had become by the interwar period.

From the professional and sociological point of view, another set of factors was at work in maintaining, even exacerbating, the widening gap between genetics and development. The success that genetics enjoyed, in particular from 1910 onward, was threatening to eclipse other more traditional areas of biology, such as embryology. In the USA for example, through the 1910s and 1920s, the work of Morgan and his group at Columbia, R. A. Emerson at Cornell, Frank Zeleny at the University of Illinois, E. B. Babcock and his students at the University of California, Berkeley, E. M. East and W. E. Castle at Harvard, Donald F. Jones at the Connecticut Agricultural Station, H. J. Muller at Texas, and L. J. Stadler at the University of Missouri (to name only a few) had made remarkable progress in elucidating the arrangement of genes on chromosomes, patterns of crossing-over and genetic exchange, and the nature of mutagenesis. Not only were geneticists making headway with problems that only a decade or two earlier seemed inexplicably complex, but Mendelism, in conjunction with the chromosome theory, seemed to fulfill all the requirements of the "new" biology: it was mechanistic, quantitative, and experimental; and its hypotheses could be tested and confirmed or rejected in a manner akin to that practiced in the physical sciences. Moreover, by restricting its questions to mechanisms of transmission, genetics became a "do-able" science. Conclusions were no longer a matter of speculation and metaphysical postulates.

The phenomenal success of classical genetics up through the 1950s thus gave it a prestige and power that embryology, for all its many advances, lacked. In the "struggle for authority," as Jan Sapp (following Pierre Bourdieu) has emphasized, scientists compete for funds, graduate students, space, access to journals and meeting programs, and cultural hegemony within and between their disciplines.[36] For reasons not unrelated to the transcendence of genetics over embryology, focus on the cell nucleus also led to ignoring the cytoplasm, which embryologists were convinced played a key role in early, if not all, stages of differentiation. This point is well illustrated by Morgan's claim in 1926 that "In a word, the cytoplasm can be ignored genetically."[37] (The claim carries a certain irony, since Morgan himself objected to Mendelism for a decade on the grounds that, among other issues, it ignored the cytoplasm and the role it obviously played, from the embryologist's perspective, in the early stages of development.)

The dominance of genetics over embryology was more than intellectual, however. This is again reflected in institutional ways at the CIW. The budget of the Department of Embryology in 1920 was $43,128, while that of its Department of Genetics (at Cold Spring Harbor), was $78,343.[38] By 1928 the differential was even greater: $16,551 for Embryology, and $211,203 for genetics.[39] As I have suggested elsewhere, much of this funding differential appears to have resulted from the role genetics was seen as playing in agriculture (for many of the same reasons that modern genomics is so highly underwritten today).[40] The financial disparity could not help but make embryologists feel to some extent a poor second cousin to their brilliant, if younger, disciplinary cousin.

This disciplinary boundary concern was perhaps nowhere more clearly, and explicitly, stated than by Ross G. Harrison, in his address as retiring vice-president of the Zoology section of the American Association for the Advancement of Science in December, 1936. The address is noteworthy in its balanced view of genetics from the embryologist's point of view. Beginning with generous praise for recent advances in genetics, Harrison goes on to warn of geneticists' potential for colonial expansion into embryology:

> The location of genes in the chromosomes, the proof of their linear order, the association of somatic characters with definite points in the chromosomes, in short, the whole development of the gene theory is one of the most spectacular and amazing achievements of biology in our times. The liaison between genetics and embryology is now established . . . However, . . . the predicted gold rush to our own territory is upon us and times are strenuous again . . . Now that the necessity of relating the data of genetics to embryology is generally recognized and the 'Wanderlust' of geneticists is beginning to urge them in our direction, it may not be inappropriate to point a danger in this threatened invasion. The prestige of success enjoyed by the gene theory might easily become a hindrance to the understanding of development by directing our attention solely to the genom [sic], whereas cell movements, differentiation and in fact all developmental processes are actually affected by the cytoplasm. Already we have theories that refer the processes of development to genic action and regard the whole performance as no more than the realization of the potencies of the genes. Such theories are altogether too one-sided.[41]

Thus, embryologists were not only bothered by the rapid progress geneticists had made in elucidating the patterns of transmission in a wide variety of organisms and the broad general principles that were emerging as a result.

They also feared for the very existence of their disciplinary boundaries, as geneticists, flushed with the excitement (and to some extent arrogance) of success, were proposing to enter embryology and reduce it to genetics – that is, to subsume the fundamental problems of embryology under the problem of differential gene action. In the eyes of embryologists, this would not be a true synthesis, but only a restriction of their work to a sub-field of genetics. Such feelings, coming from both camps, could not but make a true synthesis more difficult to effect.

Harrison need not have worried. The "threatened liaison" was not quite so far along as he imagined. It did come, though in different ways, beginning in the 1960s with the expansion of molecular genetics, first in the form of the operon concept, then with the discovery of promoter sites, transcription factors, and a large assembly of genetic control elements that suggested at least how development might be controlled at the level of the gene. Today, we are perhaps more on the road to a true synthesis, institutionally as well as intellectually, as positions in the field of evolution, genetics, and development ("evo–devo" in the latest lingo) are becoming increasingly more common. Some of the molecular biology being pursued at the present CIW Department of Embryology, described in Donald Brown's and Allan Spradling's chapters, chapters 7 and 8, is indicative of how much the fields of genetics and embryology have now begun to merge, or at least to find more common ground. The biological community has come a long way from Haeckel's and Weismann's original "evo–devo" programs. But the surface has only been scratched. In the twenty-first century it is likely to be the unified paradigm of heredity–development–evolution, understood starting at the molecular level, and integrated into ever-higher levels of organization, that will become one of the most profound and exciting areas of investigation in the life sciences.

Acknowledgements

In preparing this chapter I was fortunate to have the help of the staff at the Carnegie Institution of Washington, in particular, Margaret Hindle Hazen, Sharon Bassin, and Tina McDowell. The CIW archives were an important source of some details regarding the management of the Department of Genetics, and I would like to thank the archivist for guidance and for basically letting me peruse the various holdings informally. As usual, Jane Maienschein has provided very helpful, critical commentary that has improved the chapter immensely. Marie Glitz has been very supportive as well, and has helped keep my formatting on track! Don Brown and Alan Spradling offered very helpful comments on the earlier version of this chapter, emphasizing the need to tie the history of the Department of Genetics more closely to that of the Department of Embryology.

Notes

1. Scott F. Gilbert, "The embryological origins of the gene theory," *Journal of the History of Biology* 11 (1978), pp. 307–51.
2. See, for example, Stephen Jay Gould, *Ontogeny and Phylogeny* (Cambridge, MA: Harvard University Press, 1977); and Frederick B. Churchill, "Believing in the biogenetic law: the internal critique." Paper presented at the Dibner Institute Seminar, *From Embryology to Evo–Devo*, 18–19 October 2002.
3. Garland E. Allen, "Heredity under an embryological paradigm: the case of genetics and embryology," *Biological Bulletin* 168 (1985), pp. 107–21.
4. Thomas Hunt Morgan, The Embryology and Phylogeny of the Pycnogonids, Ph.D. Dissertation, 1891, Johns Hopkins University; Charles B. Davenport, "Observations on budding in *Paludicella* and some other bryozoans," *Bulletin of the Museum of Comparative Zoology at Harvard College* 22 (No. 1, 1891).
5. Gould, *Ontogeny and Phylogeny*; Garland E. Allen, *Life Science in the Twentieth Century* (New York: Cambridge University Press, 1978), especially chapter 2; Nicholas Rasmussen, "The decline of recapitulationism in early twentieth-century biology: disciplinary conflict and consensus on the battleground of theory," *Journal of the History of Biology* 24 (1991), pp. 51–89.
6. See, for example, Allen, *Life Science*, especially chapters 1–3; for a critique of the "revolt from morphology" thesis, see the special section, "American morphology at the turn of the century," with an introductory essay, "Were American morphologists in revolt?" by Jane Maienschein, Ronald Rainger, and Keith R. Benson, *Journal of the History of Biology* 14 (1981), pp. 89–114; Ron Rainger, Keith R. Benson, and Jane Maienschein (eds.), *The American Development of Biology* (Philadelphia, PA: University of Pennsylvania Press, 1988); and Jane Maienschein, *Transforming Traditions in American Biology* (Baltimore, MD: Johns Hopkins University Press, 1991); the varieties of morphology have been discussed by Lynn K. Nyhart, "Learning from history: morphology's challenges in Germany ca. 1900," *Journal of Morphology* 252 (2002), pp. 2–14; Scott F. Gilbert (ed.), *A Conceptual History of Modern Embryology* (New York: Plenum Press, 1991), especially the essays by Frederick B. Churchill, Jean-Louis Fischer, and Jane Maienschein.
7. Allen, *Life Science*, chapter 2.
8. Thomas Hunt Morgan, "Sex-limited inheritance in *Drosophila*," *Science* 32 (1910), pp. 120–2.
9. Garland E. Allen, "Mendel and modern genetics: the legacy for today," *Endeavour* (June, 2003), 27: 63–8.
10. See Davenport's presidential address to the American Society of Naturalists on 19 December 1907. Reprinted as "Cooperation in research," *Science* 25 (1907), pp. 361–6.
11. Garland E. Allen, "The Eugenics Record Office at Cold Spring Harbor, 1910–1940: an Essay in institutional history," *Osiris* (second series) 2 (1986), pp. 225–64. See also Executive Committee, 3 October 1902, Carnegie Institution of Washington Archives, CIW Record Book, p. 57.
12. Charles B. Davenport and Merle Ekas, *Statistical Methods in Biology, Medicine and Psychology*, 4th edn. (New York: John Wiley and Sons, 1936).
13. E. Carlton McDowell, "Charles Benedict Davenport, 1866–1944. A study of conflicting influences," *Bios* 17 (No. 1, March, 1946), pp. 2–50, see especially p. 14.

14. Charles B. Davenport, *Experimental Morphology* (New York: Macmillan, 1897, 2nd edn., 1899).

15. Barbara A. Kimmelman, "The American Breeders Association: genetics and eugenics in an agricultural context, 1903–1913," *Social Studies of Science* 13 (1983), pp. 163–204.

16. Ibid. See also any of the *American Breeders Association Reports* from the first volume in 1903 (it changed its name in 1910 to the *American Breeders Magazine* until it ceased publication in 1917). Ironically, it was in the *ABA Report* of 1909 that T. H. Morgan launched one of his most vociferous criticisms of mendelian theory: T. H. Morgan, "What are factors in mendelian inheritance?," *American Breeders Association Report* 5 (1909), pp. 365–8.

17. See William Provine, *The Origins of Theoretical Population Genetics*, (Chicago, IL: University of Chicago Press, 1971, revised edn., 2001), especially chapters 2 and 3.

18. McDowell, "Charles Benedict Davenport," pp. 22–3.

19. Ibid., p. 18.

20. Davenport to Billings, quoted in McDowell, "Charles Benedict Davenport," p. 19.

21. See Minutes of the CIW Executive Committee, 3 October 1902 and 12 December 1905, CIW Archives, Record Book, pp. 57, 468–75; see also Bentley Glass, "The strange encounter of Luther Burbank and George Harrison Shull," *Proceedings of the American Philosophical Society* 124 (1980), pp. 133–53.

22. Minutes of the Executive Committee, 3 October 1903, CIW Archives, Record Book, p. 56.

23. McDowell, "Charles Benedict Davenport," p. 24.

24. Ibid., pp. 24–25; for details on McDowell's own work, see Philip Pauly, "How did the effects of alcohol on reproduction become scientifically uninteresting?," *Journal of the History of Biology* 29 (1996), pp. 1–28.

25. Charles B. Davenport and Gertrude C. Davenport, "Heredity of eye-color in man," *Science* 26 (1907), pp. 589–92.

26. Allen, "The Eugenics Record Office at Cold Spring Harbor," pp. 225–64.

27. Allen, "The Eugenics Record Office at Cold Spring Harbor," p. 236.

28. Merriam, memo 12/31/28, p. 1.

29. Ibid., p. 4.

30. Merriam, memo 10/12–13/23, CIW Archives, Merriam Folder.

31. Merriam, memo 11/10–14/23, pp. 2–3, CIW Archives, Merriam folder.

32. Merriam, memos 3/20/29 and 3/23/29, CIW Archives, Merriam folder.

33. Merriam, memo 11/10/23, CIW Archives, Merriam folder.

34. Merriam, memo 10/14/29, CIW Archives, Merriam folder.

35. For details on McClintock, see Nathaniel C. Comfort, *The Tangled Field* (Cambridge, MA: Harvard University Press, 2001); and also the earlier study by Evelyn Fox Keller, *A Feeling for the Organism* (New York: W. H. Freeman, 1983).

36. Jan Sapp, *Beyond the Gene: Cytoplasmic Inheritance and the Struggle for Authority in Genetics* (New York: Oxford University Press, 1987), pp. xiv–xvi; Pierre Bourdieu, "The specificity of the scientific field and the social conditions of the progress of reason," *Social Science Information* 1 (1975), pp. 19–47.

37. Thomas Hunt Morgan, "Genetics and the physiology of development," *American Naturalist* 60 (1926), pp. 489–515.

38. Carnegie Institution of Washington *Year Book* 20 (1920–1), p. 18.

39. Carnegie Institution of Washington *Year Book* 27 (1927–8), p. 18.

40. Garland E. Allen, "The reception of mendelism in the United States, 1900–1930," *Comptes Rendu l'Academie des Sciences, Paris, Sciences de la Vie* 323 (2000), pp. 1081–8.
41. Ross G. Harrison, "Embryology and its relations," *Science* 85 (1937), pp. 369–74.

Bibliography

Allen, Garland E., *Life Science in the Twentieth Century*, revised edn. (New York: Cambridge University Press, 1978).

"Heredity under an embryological paradigm: the case of genetics and embryology,"*Biological Bulletin* 168 (1985), pp. 107–21.

"The Eugenics Record Office at Cold Spring Harbor, 1910–1940: an essay in institutional history," *Osiris* (second series) 2 (1986), pp. 225–64.

"The reception of mendelism in the United States, 1900–1930," *Comptes Rendu de l'Academie des Sciences, Paris, Sciences de la Vie* 323 (2000), pp. 1081–8.

"Mendel and modern genetics: the legacy for today,"*Endeavour* (June, 2003), 27: 63–8.

Bourdieu, Pierre, "The specificity of the scientific field and the social conditions of the progress of reason," *Social Science Information* 1 (1975), pp. 19–47.

Carnegie Institution of Washington *Year Book* 20 (1920–1).

Year Book 27 (1927–8).

Churchill, Frederick B., "Believing in the biogenetic law: the internal critique." Paper presented at the Dibner Institute Seminar, *From Embryology to Evo–Devo*, 18–19 October 2002.

Comfort, Nathaniel C., *The Tangled Field* (Cambridge, MA: Harvard University Press, 2001).

Davenport, Charles B., *Experimental Morphology* (New York: Macmillan, 1897, 2nd edn., 1899).

"Observations on budding in *Paludicella* and some other bryozoans," *Bulletin on the Museum of Comparative Zoology at Harvard College* 22 (No. 1, 1891).

"Cooperation in research," *Science* 25 (1907), pp. 361–6.

Davenport, Charles B. and Gertrude C. Davenport, "Heredity of eye-color in man," *Science* 26 (1907), pp. 589–92.

Davenport, Charles B. and Merle P. Ekas, *Statistical Methods in Biology, Medicine and Psychology*, 4th edn. (New York: John Wiley, 1936).

Gilbert, Scott, F., "The embryological origins of the gene theory," *Journal of the History of Biology* 11 (1978), pp. 307–51.

(ed.), *A Conceptual History of Modern Embryology* Volume 7 of *Developmental Biology, A Comprehensive Synthesis* (New York: Plenum Press, 1991).

Glass, Bentley, "The strange encounter of Luther Burbank and George Harrison Shull," *Proceedings of the American Philosophical Society* 124 (1980), pp. 133–53.

Gould, Stephen Jay, *Ontogeny and Phylogeny* (Cambridge, MA: Harvard University Press, 1977).

Haeckel, Ernst, *Generelle Morphologie der Organismen: Allgemeine Grundzüge der organischen Formen-Wissenschaft, mechanische begründet durch die von Charles Darwin reformirte Descendenz-Theorie*, 2 vols. (Berlin: Georg Reimer, 1866).

Harrison, Ross G., "Embryology and its relations," *Science* 85 (1937), pp. 369–74.

Keller, Evelyn Fox, *A Feeling for the Organism* (New York: W. H. Freeman, 1983).

Kimmelman, Barbara A., "The American Breeders Association: genetics and eugenics in an agricultural context, 1903–1913," *Social Studies of Science* 13 (1983), pp. 163–204.

McDowell, E. Carlton, "Charles Benedict Davenport, 1866–1944. A study of conflicting influences," *Bios* 17 (No. 1, March, 1946), pp. 2–50.

Maienschein, Jane, *Transforming Traditions in American Biology, 1880–1915* (Baltimore, MD: Johns Hopkins University Press, 1991).

Maienschein, Jane, Ronald Rainger, and Keith R. Benson, "Introduction" to "American Morphology at the turn of the Century," Special section of the *Journal of the History of Biology* 14 (1981), pp. 83–158.

Morgan, Thomas Hunt, "The Embryology and Phylogeny of the Pycnogonids," Ph.D. Dissertation, 1891, Johns Hopkins University.

"What are Factors in Mendelian Inheritance?," *American Breeders Association Report* 5 (1909), pp. 365–8.

"Sex-limited inheritance in *Drosophila*," *Science* 32 (1910), pp. 120–2.

"Genetics and the physiology of development," *American Naturalist* 60 (1926), pp. 489–515.

Nyhart, Lynn K., "Learning from history: morphology's challenges in Germany ca. 1900," *Journal of Morphology* 252 (2002), pp. 2–14.

Pauly, Philip J., "How did the effects of alcohol on reproduction become scientifically uninteresting?," *Journal of the History of Biology* 29 (1996), pp. 1–28.

Provine, William, *The Origins of Theoretical Population Genetics* (Chicago, IL: University of Chicago Press, 1971, revised edn., 2001).

Rainger, Ron, Keith R. Benson, and Jane Maienschein (eds.), *The American Development of Biology* (Philadelphia, PA: University of Pennsylvania Press, 1988).

Nicholas Rasmussen, "The decline of recapitulationism in early twentieth-century biology: disciplinary conflict and consensus on the battleground of theory," *Journal of the History of Biology* 24 (1991), pp. 51–89.

Sapp, Jan, *Beyond the Gene: Cytoplasmic Inheritance and the Struggle for Authority in Genetics* (New York: Oxford University Press, 1987).

Weismann, August, *Das Keinplasm. Eine Theorie den Verenburg* (Jena: Gustav Fisher, 1892).

7

THE DEPARTMENT IN THE SECOND HALF OF THE TWENTIETH CENTURY

DONALD D. BROWN

Department of Embryology, Carnegie Institution of Washington

In his final annual report as Director of the Department of Embryology in 1955, George Corner wrote "The great strength of the Department has always been the association and interplay of morphological and physiological thinking". However, "a gradual shift of emphasis was necessary."[1] This would involve "physiology, biophysics and biochemistry. The Department of Embryology will of necessity have to consider, as far as its means and staff can reach, the full gamut of life structure from the electron to the whole organism." These comments signaled the end of an era in which the Department excelled at anatomy and reproductive biology. With Corner about to retire, the Carnegie Institution of Washington's new President, Caryl Haskins, was faced with the decision that has confronted the Institution each time there is a major change in leadership and direction of a department. Should the Department of Embryology be disbanded and resources directed to a completely new area of research? If the field of embryology had a future worthy of the Institution's continued support, could a small department play an important role, presumably in directions predicted by George Corner?

James David Ebert, the Fourth Director

No doubt the decision to continue the Department of Embryology relied in part on finding a new Director who could lead the Department into new and exciting directions. There were no candidates from within the existing staff. Caryl Haskins, the new President, chose James David Ebert, a Professor of Biology at Indiana University, as Director of Embryology. Corner himself had recommended Ebert to Haskins, but the major enthusiasm for Ebert's appointment came from Benjamin Willier, then the Chairman of the distinguished Biology Department at Johns Hopkins University. Willier had been Ebert's thesis advisor and had watched Ebert's subsequent rapid rise through the academic ranks, first at MIT and then at the University of Indiana. Ebert was 34 years old when he began his tenure as the fourth Director of the Department of Embryology on January 1, 1956. Haskins and Ebert had decided that a new direction for the Department would require a modern state-of-the art facility. For the previous 40 years the Department had had a natural alliance with the Departments of Anatomy, Obstetrics, and Pathology at its location on the campus of the Johns Hopkins Medical School. However, Ebert had no constraints on the site of the new building. Indeed, there were universities in other cities that had shown interest in hosting the Department on their campus. No doubt, Ebert's close relationship with Willier, plus the fact that he had already received a joint appointment in Biology at Johns Hopkins and had begun to accept their graduate students soon after his arrival, played major roles in his decision. He knew that the Department's research would evolve toward areas that were represented more closely within the Hopkins Biology Department. Soon after Ebert's arrival the Institution negotiated a long-term lease with Johns Hopkins to occupy a one acre wooded property at the northwest corner of the Homewood undergraduate campus located five miles across town from the Medical School. Johns Hopkins University would own both the land and the building. The Department of Embryology would have sole use of the structure and grounds and the responsibility of its cost and upkeep. Carnegie would pay Hopkins one dollar a year for ninety-nine years, after which Johns Hopkins could use the building and land for whatever purpose they might choose.

Ebert hired a Boston architectural firm; however, the plans required the approval of the Hopkins Board, and all buildings on the Homewood campus at that time were colonial in design. The modern appearance of the proposed Embryology structure generated immediate opposition at the highest levels of the University (Fig. 7.1). Ebert pointed out that the proposed building could not even be seen from the main campus. Finally, the Hopkins Trustees agreed. Subsequently, other structures in the same style were built around the periphery of the Hopkins campus. Unlike other Hopkins-related buildings, the Department was directly across the street from one of the oldest high-rise apartments in the city. Elizabeth Ramsey's monkey colony was housed with an outside run that faced the apartments. The neighbors complained soon after the Department moved in about the antics of the monkeys. Subsequently, we learned that they could only see the monkeys with binoculars. The monkey colony was discontinued upon the retirement of Elizabeth Ramsey in 1971 (See Hanson, chapter 3 this volume). This colony, which had been the source of so many discoveries in reproductive biology, would be impossible today for our Department to maintain and protect, considering the level of opposition that now exists toward primate research.

Figure 7.1 The Department of Embryology building located on the northwestmost peninsula of the Johns Hopkins University Homewood campus viewed from San Martin Drive.

The plan of the new laboratory predicted some fundamental changes in our future style of research and training. Although it was designed to house eight independent faculty members called "staff members," the same number as in the old building, there were also labs for fellows, graduate students, and sabbatical scientists. This anticipated a departure from the previous forty-five years, when individual staff members worked independently, generally with one technician and perhaps a visiting scientist. There had been no tradition of training graduate students or postdoctoral fellows in the Department when it was located at the Medical School. However, the Department had always hosted temporary independent scientists called "fellows" to study the human embryo collection or work with particular faculty members. In addition, the Department of Embryology had been an attractive escape for academics on sabbatical who wanted to pursue their particular interests for a year unencumbered by students or teaching.

The Director of the Department in those days made all administrative decisions. Although staff members carried out independent research, the Director appointed them in the first place. Since CIW faculty has never had traditional academic tenure or even a written contract, the staff members served at the pleasure of the Director. Therefore, Ebert was totally responsible not only for the new building but also for the faculty appointments for the first ten years that he was Director. Sometime in the mid 1960s, Ebert began to consult with his faculty on new appointments. For a decade, Ebert wrote the entire detailed yearly departmental research summary for the annual CIW *Year Book*. In addition, the Director had full control of the budget, since all of the money came from CIW endowment until 1975, when Igor Dawid and I applied to the National Institutes of Health for the Institution's first extramural grants. These outside grants were awarded one year before Ebert stepped down from the directorship.

· Trained in experimental embryology with Willier, Ebert used immunological tools to study the development of the chick embryo. He was especially fascinated with the "graft versus host response," a phenomenon that still interests the immunological community. When foreign tissue is transplanted to the chorio allantoic membrane of the chick embryo it results in a huge enlargement of the host spleen. Ebert and his colleagues recognized that this specific reaction to a foreign transplantation had an immunological explanation. He and his colleagues applied many experimental parameters to this system.[2]

Ebert's main interests, after assuming the directorship and moving the Department to the Johns Hopkins Homewood campus, turned toward building a strong Department of independent basic scientists and participating in national and international science policy. The Department moved to its new location in August, 1961. When Ebert relinquished the directorship in 1976 he had overseen a change in the research orientation of the Department that was even more profound than had been forecast by George Corner. During his career, Ebert was President of eight professional societies and trustee of thirteen national and international organizations, and he held appointments to some forty different advisory boards and committees of government agencies, and universities. At one time in his career, Ebert held the presidency or was on the board of every national and international society concerned with developmental biology. He was elected for two terms (twelve years) as Vice President of the National Academy of Sciences. Ebert had a long and close association with the Marine Biological Laboratory (MBL). He directed the embryology course in the early 1960s and became President of MBL from 1970 to 1978. Throughout his career Jim Ebert was especially concerned with promoting scientific interactions with Japan. Over the past forty years more than thirty Japanese scientists have spent time in the Department of Embryology as graduate students, postdoctoral fellows, or sabbatical workers, including some of today's most distinguished Japanese scientists. Ebert traveled to Japan often to serve on advisory boards. More recently, he developed a similar relationship with Chinese scientists. Ebert succeeded Philip Abelson as President of the CIW, serving from 1978 to 1987.

Three of the staff members appointed by George Corner remained when Ebert became Director, including physiologists David Bishop and Bent Boving. Bishop was an expert on sperm motility. The rhythmic movement of sperm has many of the features of muscle contraction. Injection of guinea pigs with an extract from their own testes destroys their sperm. This is an autoimmune response against some specific component in testes that David Bishop spent many years trying to identify.[3] Bishop retired in 1967. Boving sought to understand the precise equal spacing of the multiple embryos that implant along the side of the rabbit uterus.[4] The process of implantation requires a receptive uterine wall, an adhesive interaction between the wall and the blastocyst, and, finally, invasion of the trophoblast into the uterine wall. Boving moved to the Department of Anatomy of Wayne State University Medical School in 1968. The third long-time staff member, Robert Burns, was a distinguished reproductive biologist and a member of the National Academy of Sciences. Burns studied opossums, which are born at an early stage of their development. They crawl into the mother's pouch to complete development, thus making them available for analysis. Amongst other discoveries, Burns found the conditions to reverse the sex of the developing opossum with hormones.[5] Burns retired from the Department in 1962. This era of reproductive physiology is elaborated on in greater detail by Adele Clark, chapter 4, this volume.

One of Ebert's first acts as Director was to promote Elizabeth Ramsey to staff member. Corner had appointed Ramsey as a "research associate" a decade earlier. By the time of Ebert's appointment, Ramsey's interest in human embryology had changed completely to the physiology of the primate placenta. Nevertheless, she remained the unofficial curator of the human embryo collection until her retirement. In this volume, Adrianne Noe, chapter 2, describes the fate of the human embryo collection and its current home in Washington, DC.

In the late 1950s, Ebert appointed two physiologists to the staff. Royal (Bud) Ruth had much the same research interests in developmental immunology as Ebert, and they collaborated during Ruth's brief time in the Department. Ruth moved to Canada in 1961. Robert DeHaan began a distinguished career studying the formation of the chick heart. The progenitor cells for the heart are formed on either side of the embryo and then migrate together to form the heart rudiment. The problem of morphogenetic movements, or the remarkable journey that some cells take to their final location in the embryo, was and still remains an essential problem in embryology. DeHaan used pharmacological inhibitors to deflect the route of these cells. He learned how to identify the cells that would give rise to the heart and how to inhibit their migration. During this era, the use of pharmacological inhibitors was popular as a means to study function.[6] DeHaan moved to Emory University as a Professor of Anatomy in 1975.

In the mid 1950s, the major figures in the field of embryology studied the interaction of tissues as cells and organs become specified during development. The principal research tool was the surgical transplantation of tissues from one site of the embryo to another. The founder of this school called "experimental embryology" was Hans Spemann, who won the Nobel Prize in 1935 for his discovery of the embryonic organizer. This was an approach and set of questions that would be studied again decades later with new powerful genetic and molecular techniques. Some tissues induce the development of others at very specific times of development. The rules and features of embryonic "determination" were being defined. What is the information provided by one tissue to another that decides that tissue's developmental fate? When during development did the induction of one type cell's fate by another take place and how could it be assayed? In 1957, Ebert appointed Mary Rawles, whom he had known from his graduate student days at Johns Hopkins. She had been a research associate for many years with Benjamin Willier and was skilled at dissecting and transplanting chick embryo tissues. Rawles studied the interaction of layers of the skin. She found that the underlying dermis instructs the epidermis on its final fate. Three different epidermal choices that she investigated were feathers, scales, and beak. The receptivity of the epidermis to instruction by the dermis was restricted to early developmental stages.[7] Mary Rawles retired in 1966.

Although the distinguished embryologists in the early 1900s recognized the importance of the nucleus and chromosomes in heredity, the discipline of genetics played no part in experimental embryology in the first half of the twentieth century. The organisms that were popular for experimental embryology were chickens, frogs, and sea urchins, none of which were suitable for traditional Mendelian genetics. The rules of genetics were not being applied to development, probably because developmental processes seemed much too complicated. Popular genetic model systems up to 1960 were the single-cell bacteria *Escherichia coli*, yeast, and *Neurospora*, all organisms that do not undergo development. Many of the rules of inheritance were learned from *Drosophila* (the fruit fly), and the suitability of the mouse for genetics had been established. However, neither of these latter two eukaryotic organisms was a popular model system for the embryologists of that era. The same organisms that were popular with experimental embryologists were used for biochemistry because the embryos were large and plentiful. They could be synchronized in their development and raised in the laboratory. Biochemistry became the dominant biological discipline of the 1950s and 1960s. An especially popular branch was enzymology, the purification and characterization of enzymes. Occasionally, a biochemist would become interested in a developmental phenomenon. This generally meant measuring the activity of an enzyme at different developmental stages or as a response to some perturbation such as wounding, regeneration, or pharmacological interference.

When I was an editor for the *Journal of Biological Chemistry* in the 1960s, I received two types of articles to review, those involving the activity of an enzyme measured as it changed during the development of an embryo, and the analyses of histones. For some reason, these nuclear proteins were considered to be in the province of development. The biochemist's tradition was to purify molecules. They were not ready to come to grips with complex interacting "dirty" systems such as a developing embryo.

The insularity of embryology (and many other fields) during that era can hardly be overestimated. Comparative embryology courses were taught in most biology departments as a purely descriptive set of anatomical facts. Medical students spent a few days in their anatomy course learning the outlines of human embryology. Of course, the textbook pictures and description of human embryos came from the Carnegie Collection. Clinical obstetricians had an interest in development, but there were no findings in the field that impacted the clinics. The study of birth defects, teratology, was phenomenological. Research involved applying substances to pregnant rodents or to embryos of other species and then cataloguing the defects. Experimental embryologists trained other experimental embryologists, and they used a set of model organisms that were not popular with other biological disciplines.

The biochemists (molecular biologists)

In 1960, the last year of the Department in the New Hunterian building at Johns Hopkins Medical School, a visit to the Department of Embryology gave the impression of an anatomy department or perhaps a museum devoted to human embryology. There were plaster models of human embryos filling all available shelves and tables. The major storage room contained the accumulated slides of human embryos from fifty years of collecting. The staff lunched together in a library that had more plaster models of embryos on the shelves than books. Although none of the faculty studied the human embryo, there was always a visiting scientist or two using the collection. There were extensive monkey-room facilities with an expert staff of animal technicians, and an operating room used for Elizabeth Ramsey's experiments on the placenta of the monkey. By contrast, one ultracentrifuge and one manual gas flow radioactive counter represented the entire departmental inventory of biochemical equipment. I arrived in the Fall of 1960 as a student with a background in biochemistry eager to learn and study embryology. I had just spent one year of postdoctoral training with Jacques Monod at the Pasteur Institute in Paris. The year before my arrival in Paris, Pardee, Jacob, and Monod had reported one of the great discoveries of modern biology. They had found genetic evidence that the action of genes in the bacterium *E. coli* is controlled by the product of another gene called a repressor. This provided the first glimpse into the molecular basis of the control of gene expression, a subject at the heart of embryonic development. I left Paris for Baltimore fully indoctrinated with this and other marvelous discoveries that were being made using bacterial genetics and biochemistry. This combination of fields was soon to be called "molecular biology."

The control of gene expression was on the minds of developmental biologists. The term that was used by embryologists to explain development was "differential gene expression." By 1960, it was known that RNA was the intermediate between the genes (DNA) and protein. Popular biochemical analyses of embryonic development measured the change in some protein, usually an enzyme. Proteins, the ultimate products of genes, were indirect indicators of gene activity, while a direct assessment of gene activity measured RNA, the gene product. By 1960, there were three recognized kinds of RNA. I measured the amount of each of the three kinds of RNA at different developmental stages of *Rana pipiens* embryogenesis and found that they were under different control. Soon after this discovery, I read about the remarkable "anucleolate" mutant that had been described for the amphibian *Xenopus laevis* in the Zoological Laboratories of Oxford University. John Gurdon and I proved that the defect was the inability of mutant embryos to synthesize one of these classes of RNA, ribosomal RNA.[8] Subsequently, Max Birnstiel in Edinburgh demonstrated that the mutation was a deletion of the ribosomal RNA genes. In 1966, Birnstiel and his colleagues developed a biochemical method to isolate the ribosomal RNA genes from the bulk of the genomic DNA.[9] This was the first isolation of a gene, and it was carried out ten years before the era of DNA cloning. The study of specific genes and their RNA products was a major research focus of the biochemists in the Department from 1964 to 1975 when the recombinant DNA period began.

In 1963, Igor Dawid joined the Department as an independent fellow and became a staff member in 1966. With training in biochemistry at the University of Vienna and then advanced studies at MIT, Dawid tackled the long-unexplained observation that frog eggs contain very large amounts of DNA. The egg is just a single cell and should, therefore, have vanishingly small amounts of DNA. Dawid purified and quantified the DNA in *Xenopus* eggs.[10] His experiments coincided with other research, the first discovery of DNA in the mitochondria. It was known that frog eggs have abundant stores of mitochondria. Soon, Dawid had confirmed that the excess frog egg DNA was located not in the nucleus but in the mitochondria. This began a series of experiments in which he characterized the mitochondrial DNA and nuclear DNA during early development. His interest in DNA and my interest in a specific set of genes caused us to work together for a decade. Our first collaboration proved that the ribosomal RNA genes are amplified specifically in the oocytes of *Xenopus* and other amphibians.[11] Joseph Gall, who at that time was a Professor at Yale but who would join our Department years later, made this observation independently.[12] There are thousands of extra copies

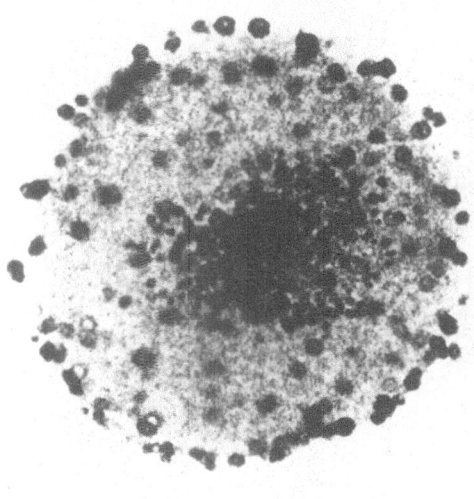

Figure 7.2 The large nucleus of a *Xenopus laevis* oocyte. The nucleus is tetraploid and should have just four nucleoli (the black bodies). However, there are several thousand in each nucleus.

of ribosomal RNA genes in the nuclei of these huge cells (Fig. 7.2). Gene amplification explained how a single cell can accumulate such a large amount of one gene product, ribosomal RNA. Amplified ribosomal DNA provided a source of purified genes that we used to study gene structure and function. Our gift of *X. laevis* ribosomal DNA to John Morrow at Stanford University led to the first cloning experiment of a gene from a higher organism into the bacterium *E. coli.*[13] This experiment initiated the recombinant DNA era, the most significant advance in genetic research since the discovery of the double helix in 1953.

During that time, we were joined by several excellent postdoctoral fellows. One of them, Ronald Reeder, addressed the accurate transcription of purified ribosomal RNA genes *in vitro*. This was the first attempt to reconstruct gene control in a test tube. Reeder became a staff member in 1968. During his ten years as a staff member Reeder introduced modern recombinant DNA methods into the Department.[14] Another postdoctoral fellow, Peter Wellauer from Switzerland, was an expert electron microscopist. He and Dawid pioneered an electron microscopic method for analyzing the structure of ribosomal RNA and its genes by visualizing the characteristic secondary structure of the molecules[15] (Fig. 7.3).

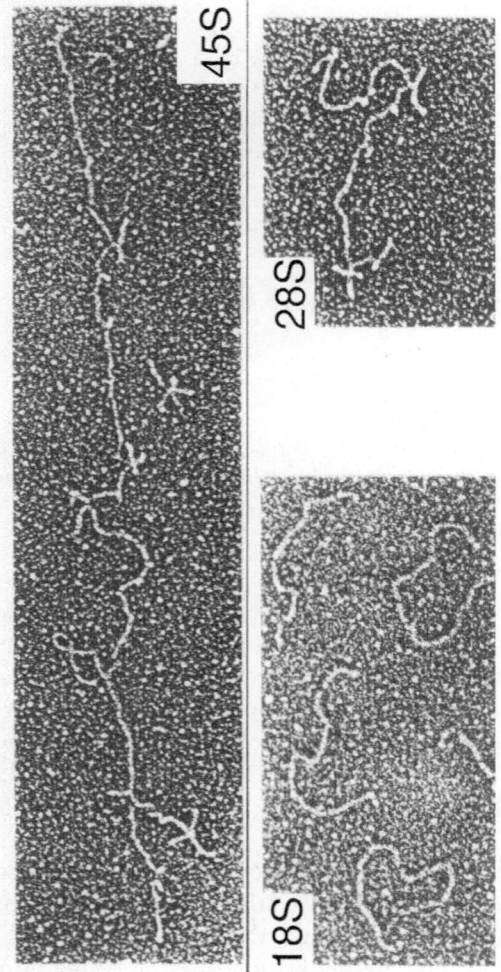

Figure 7.3 Molecules of *X. laevis* ribosomal RNA. The regions that double back on themselves are always at the same position along the molecule and therefore can be mapped precisely.

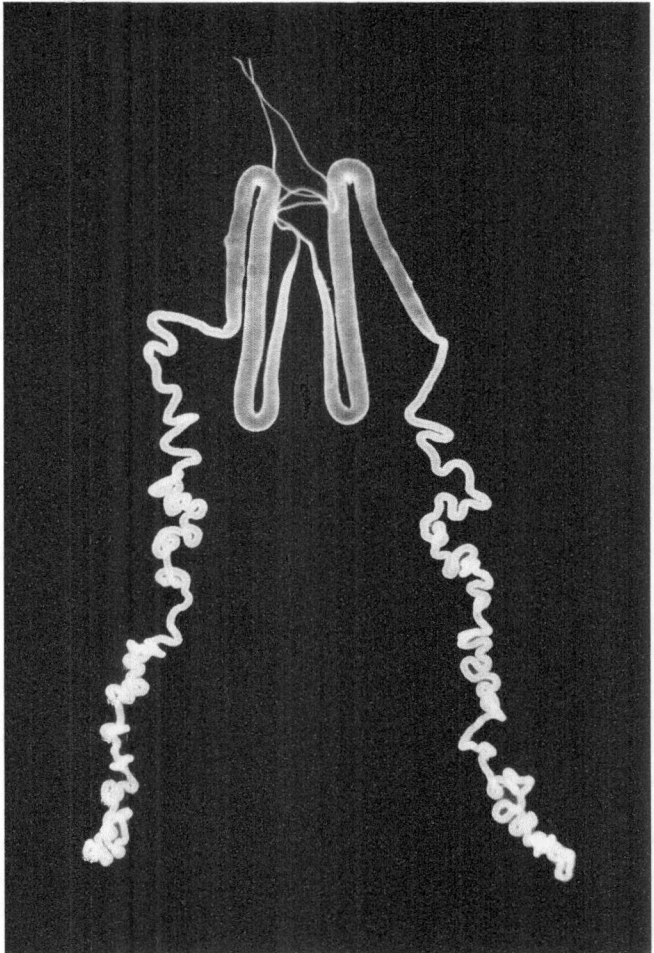

Figure 7.4 A picture of the silk gland of the Japanese silkworm, *Bombyx mori*. Silk fibroin is the major product of the posterior part of the gland, the tortuous intestine-like structure.

When amplification of the ribosomal RNA genes was discovered, it was obvious to ask whether other highly expressed genes were controlled by this same mechanism. Yoshiaki Suzuki arrived from Japan as a postdoctoral fellow in 1968 to study the silk protein of the Japanese silkworm, *Bombyx mori*. At the end of larval development the posterior region of the silk gland synthesizes a single silk protein, an example of extreme specialization (Fig. 7.4). Our object was to isolate the messenger RNA for silk protein and then to use the mRNA to determine the number of silk genes in the posterior silk gland.[16] At that time, the only mRNA for a known protein that had been identified was the mRNA encoding globin, the major protein in the red blood cell. The physical methods that had been used for the isolation of ribosomal RNA and its genes helped to accomplish both of these goals. We learned that silk gland cells could make large amounts of silk mRNA without amplifying the genes. At the time we called this "translational amplification." Each silk gene could synthesize 100,000 mRNA molecules, and each mRNA molecule in turn synthesized 10,000 protein molecules, all in the course of a few days. Suzuki was made a CIW staff member in 1973 and returned to Japan five years later to lead a research group at the new National Institute of Biology in Okazaki, Japan. He continued to study the silk system for the remainder of his scientific career and has only recently retired.

Another important gene isolated and studied in detail before the cloning era was the gene that encodes a small RNA molecule that is an integral part of the ribosome called 5S RNA. The 5S RNA genes were purified in 1970 from the DNA of *X. laevis*.[17] There is more than one kind of 5S RNA genes in amphibia, and they are developmentally regulated, making the 5S RNA gene system especially interesting for developmental questions. Over the years, we studied the structure (Fig. 7.5) and the evolution of these genes. However, our greatest interest was reserved for the control of 5S RNA synthesis. For a decade, questions about the control of gene expression could be asked uniquely with the 5S RNA gene system. By preparing specific mutations of the cloned 5S RNA gene and then assaying the ability of each mutant gene to make 5S RNA, we discovered a region in the middle of the 5S RNA gene itself that controls gene expression.[18] This "internal control region" is the binding site for a eukaryotic transcription factor called TFIIIA discovered by Robert Roeder and his colleagues. This protein was the first of a new class of proteins that bind to and control the expression of eukaryotic genes called zinc finger proteins. Our monopoly on accessible genes to study ended with the discovery of gene cloning. However, many features of the control of gene expression were explored first using the simple 5S RNA gene system.[19]

With the departures of Dawid, Reeder, and Suzuki in the late 1970s, an era of molecular biology in the Department ended. During the period from 1962 to 1980 a great deal had been learned about the structure and function of the few genes that could be isolated by physical means from genomic DNA. However, "genetics by DNA isolation" or "reverse genetics" had been pioneered using the DNA from amphibians and sea urchins, organisms that were not suitable for traditional genetics. The decade of DNA research using purified genes was replaced by the newly emerging application of genetics to development combined with the advent of powerful cloning methods that made every gene accessible for detailed study.

In 1990, my own research changed completely to the study of amphibian metamorphosis, a venerable developmental problem that had escaped the

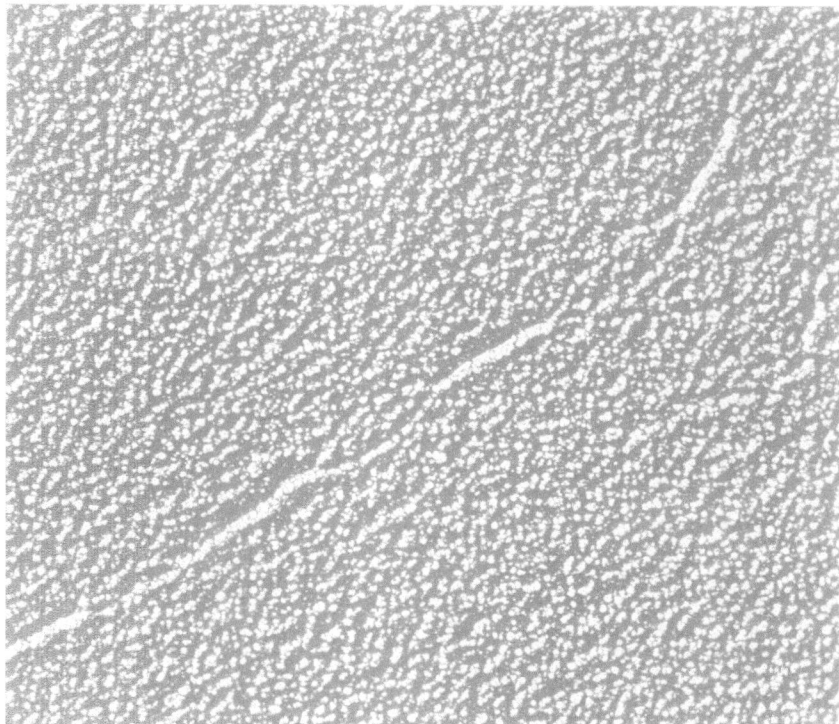

Figure 7.5 The genes that encode a small RNA called 5S RNA were purified from the genomic DNA of the frog *Xenopus laevis*. As demonstrated in this figure, many 5S RNA genes repeat along the chromosomes. The DNA has been denatured partly to show the regular repetition of gene and spacer. These genes were sequenced by Nina Fedoroff, and the purified genes were used to study gene expression in the test tube.

attention of modern biologists. The changes that occur in the tadpole as it changes into a frog are regulated by one simple molecule, thyroid hormone. The hormone works by activating or repressing different sets of genes in each responding tissue. We view it as a complex problem in the control of gene expression. This research journey has required us to develop new methods as well as to apply the full panoply of modern techniques in biology.

The cell biologists

Decades earlier the Lewises had pioneered cell culture in the Department. In 1961, Ebert appointed Irving Konigsberg as a staff member. Konigsberg had been Willier's final Ph.D. student. Following his graduation, Konigsberg had taken a position at the Aging Institute, a Baltimore branch of the National Institutes of Health. At that time there was a debate concerning the ability

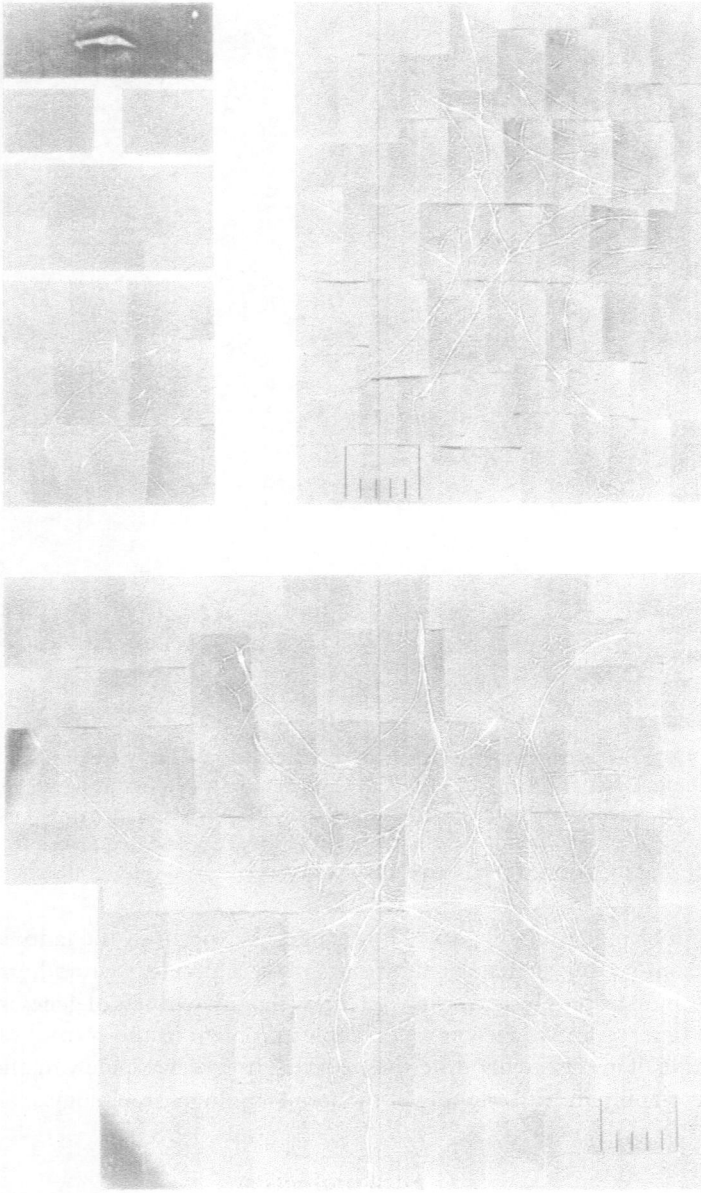

Figure 7.6 A single muscle cell, a myoblast, from an embryonic chicken was isolated in a Petri dish and allowed to divide. After several divisions the cells fused to form typical patches of muscle that could even contract.

of differentiated cells to retain their specialized function in culture. It was generally believed that specialized cells, when removed from the organism and grown in culture, would "dedifferentiate," losing their identity. Konigsberg was a meticulous cell biologist with the patience to alter methodically the culture medium to optimize conditions for explanted cells. He concentrated on chicken skeletal muscle development. Muscle has always been one of the most popular cell types for developmental biologists and biochemists to study. The myoblasts divide as single cells and then fuse together to form a multinucleated syncytium that characterizes skeletal muscle. The terminally differentiated state is easily recognized, and muscle-specific proteins had been studied for years. However, there had been no success duplicating in culture the events known to occur in the living organism. Konigsberg worked out the methods to keep muscle cells alive outside the body in culture. The secret was the use of "conditioned medium," culture medium in which other cells had grown previously. He learned that this medium could be replaced by spreading the extra-cellular protein collagen over the dishes. When he placed single myoblasts on the collagen-covered plates, they divided and then fused to form terminally differentiated muscle fibers.[20] Clearly, this finding contradicted the accepted lore, that cells in culture could not retain their specialized identity. To confirm that single cells had this capacity, he placed plastic cylinders over a single cell and watched it divide. Soon, the progeny of the single cell fused to produce multinucleated muscle fibers that were capable of contraction (Fig. 7.6). Not only did this experiment immediately change the accepted theory, but unfortunately it also caused the view to go to the other extreme. For a time, it was believed that specialized cells in culture could only give rise to cells with an identical cell fate. It was not until the modern era of molecular genetics that the idea of propagating stem cells in culture and inducing them to change their cell fate has become accepted and of great practical interest. Irwin Konigsberg played a crucial role in the history of cell culture. In 1967, Konigsberg left the Department to become a Professor of Biology at the University of Virginia.

In 1968, following the departure of Irwin Konigsberg, Douglas Fambrough was appointed to the staff. Fambrough had been trained as a biochemist at the California Institute of Technology, but his new interests lay with cell biology, especially as applied to the function of the nervous system. Specialized neuro-muscular junctions mediate the interaction between nerve and muscle. As muscle develops, it becomes innervated at these junctions, which include receptors for a crucial neurotransmitter called acetylcholine. Fambrough developed methods to study the synthesis and degradation of the acetylcholine receptors. He teamed up with a clinical neurologist at the Johns Hopkins Medical School, Daniel Drachman, to demonstrate that patients with the disorder myasthenia gravis have reduced numbers of acetylcholine receptors.[21] Drachman and Fambrough subsequently screened other muscle disorders for the integrity of their receptors. Fambrough was a distinguished neurobiologist who analyzed the most basic functions of the nervous system. He was interested in the mechanisms of nerve transmission and also studied in detail an important enzyme, sodium–potassium ATPase, whose purpose is to transmit nerve impulses by catalyzing the movement of sodium and potassium across the cell membrane. In 1985, Fambrough moved to the Johns Hopkins University as a full-time Professor in the Department of Biology.

In 1971, Richard Pagano was appointed as a staff member. Pagano had been trained as a biochemist with a strong interest in biophysical aspects of lipid behavior. He had decided to study the cell biological aspects of lipids, a subject that had been ignored in the field of biochemistry and certainly in cell and developmental biology. Lipids play a crucial structural role in the composition of all of the cell's membranes, including the two layers of the cell's plasma membrane, the Golgi apparatus, and the membranes that surround each intracellular structure. However, the details of lipid formation and then transport to the cell membrane were unknown. Pagano was also a gifted organic chemist and synthesized a variety of fluorescently tagged lipids. Pagano and his colleagues followed the entrance and exit of different lipids into either the inner or outer leaflets of the cell membrane by fluorescent microscopy.[22] In 1994, Pagano left the Department to become Professor of Biochemistry at the Mayo Clinic in Rochester, Minnesota.

Kenneth Muller, a neurophysiologist, was appointed to the staff in 1975. Muller was interested in the regeneration of nerves and the establishment of synapses between nerves. His model organism was the leech, a slimy blood-sucking creature with very large cells that can be impaled by electrodes, which has long been a model organism for neurobiologists. Muller combined the methods of cell biology with the traditional electrical recordings of neuro-physiology to follow the regeneration of severed nerves.[23] Muller left the Department in 1982 to become a Professor of Physiology and Biophysics at the University of Miami School of Medicine.

By the mid 1970s the Department hosted a distinguished and diverse cell biology group that was carrying out original research in cell biology with an emphasis on neurobiology. From 1970 to about 1980, cell biology and molecular biology were the major research interests of the Department. When I was appointed Director in 1976, the Department's faculty included the biochemists Dawid, Reeder, Suzuki, and me, the cell biologists Fambrough and Pagano, and the neurophysiologist Ken Muller. The era of developmental genetics was about to begin.

Changes made by the Fifth Director

In October 1976, Ebert resigned his directorship of the Department to become the President of the Marine Biological Laboratories (MBL) at Woods

Hole, Massachusetts. He had been associated with MBL in a leadership capacity for a long time. Philip Abelson, the CIW's President, called me to the Institution's headquarters in Washington. Although I cannot know the substance of Haskins and Ebert's original conversations leading to Ebert's appointment, they must have been very different than ours was that day. Abelson thinks deeply and is remarkably conversant with all modern fields of biology. Above all he is utterly devoid of any artifice or pretension. Since my interests were well known, there was no discussion of our philosophies or plans. He simply asked me if I would take over the directorship. Although I had had no administrative experience and had spent the previous 15 years at the bench, I had already decided to accept, mainly because I didn't want to work under anyone else. I recall trying to think of something to say other than yes. It occurred to me that I should have a salary increase so I asked him if that was in the plans. I realized immediately that this was the least appropriate question that I could have asked an academician like Abelson. It completely took him by surprise. Scientists did not discuss salary. He mumbled something to the effect that there would be an increase. I left.

I have said that Ebert made all of the decisions in the Department. While this was true for his first 10 years as Director, by the mid 1960s he had begun to consult with his faculty on appointments. By 1976, we were applying for outside grants that were destined to become the major source of our Department's money. The changes that I made in administrative policy continued an evolution toward broad participation of the faculty in all departmental decisions. There would continue to be eight faculty members, called "staff members." Although there was no formal institutional directive, each faculty member was reviewed every five years by an outside group of experts. The decision to renew a staff member remained the responsibility of the Director. The departmental policy of sharing equipment and space that I had inherited from the Ebert administration was established in the days when all financial support came from the CIW. This guiding principle has continued, even though each year a higher fraction of our budget has come from outside the Institution. I have seen so many overcrowded departments that result in fights for space. Our eight staff members have equal-sized labs, and there has always been an acceptance in our Department that the space outside our own labs is equally divisible by eight, the youngest staff member having rights to the same amount of space as the oldest. This agreement has kept even the most popular staff member's group size under control. Over the years individual staff members have averaged four to five colleagues at any time, in a ratio of about 3:1 postdoctoral fellows to graduate students.

I initiated a regular seminar series, including weekly progress reports for all graduate students and postdoctoral fellows. I instituted a weekly Monday seminar series of outside speakers and a weekly journal club in which all members of the Department participate. There has been an annual Carnegie

Mini Symposium that began the first year of my directorship. However, in the last analysis, the success of my directorship like Ebert's before me will be judged on the quality of the faculty and the importance of departmental research during the years 1976–94.

The geneticists

In 1978, a young geneticist named Samuel Ward was appointed to the Staff. Ward had carried out his thesis research at the California Institute of Technology working on bacteriophage with Bill Wood. Shortly thereafter, Wood switched to developmental research with the worm *Caenorhabditis elegans*. The value of this organism, as a result of its short life cycle and the ease with which one could generate mutations, was being recognized by the best and brightest young scientists in the USA, who in turn went to the MRC Labs in Cambridge, UK, to study with Sydney Brenner, who had introduced the organism to biologists. Following his mentor, Sam Ward changed his research to *C. elegans* by carrying out a postdoctoral fellowship at the Cambridge laboratory. He had been a junior faculty member at Harvard for several years when we recruited him to join our Department. Ward was interested in the mobile sperm of *C. elegans*, which showed a remarkable variation on normal sperm morphology. He studied developmental problems by combining genetic and molecular methods.[24]

In the late 1970s and early 1980s, many young scientists were excited at the prospect that recombinant DNA technology would finally allow the full power of genetics to be applied to the problem of development. However, such attempts remained difficult. Few genes important for development had been defined genetically, and the genes responsible for developmental events could not be cloned and characterized. Ward's group acquired and generated male sterile mutations with the expectation that some would disrupt important sperm-specific genes. Simultaneously, he isolated sperm-specific proteins, hoping that he could identify the protein affected by the mutation. For years, however, in the absence of a systematic method, all such attempts proceeded in a slow and haphazard manner. Eventually, during the 1990s, the combined power of genetics and molecular biology enabled Ward and his colleagues to identify and study some of the specific proteins in these unusual cells. Ward became Professor and Head of the Department of Molecular and Cell Biology at the University of Arizona in 1988.

Nina Fedoroff arrived at the Department as a postdoctoral fellow in 1976 at the very time that DNA sequencing was discovered. She was the first to sequence a complete gene, the repeating unit that encodes 5S ribosomal RNA. The gene sequence was done in collaboration with George Brownlee at the MRC labs in Cambridge and revealed many features of eukaryotic genes that would be found in other genomes years later when cloning genes became the standard approach. From these tandemly repeated genes came the first sequence of spacers that separate genes, and documentation of the existence of gene-like structures, which were named "pseudogenes." By comparison of the 5S RNA genes of two related species of *Xenopus*, the conservation of functional compared to nonfunctional DNA sequences was revealed, and the concept of parallel or horizontal evolution of tandem genes was established.

One day while visiting the Cold Spring Laboratories, Nina Fedoroff met Barbara McClintock. Nina learned that McClintock was the only source of a largely unpublished mass of solo research findings on the genetics of maize. Barbara McClintock was one of the world's greatest geneticists, and her career was ending as the era of molecular biology was beginning. Nina Fedoroff is a talented and brilliant molecular biologist, and she decided to apply these skills to advance with molecular methods the genetic discoveries that McClintock had made. Of the many brilliant genetic discoveries made by McClintock, perhaps the most famous was her conclusion from purely genetic data that genes can move from one part of the genome to another. Fedoroff set out to identify these mobile genetic elements in maize and in a few short years had done so.[25] She was appointed to the faculty in 1978 and within just a few years became one of the most distinguished plant geneticists in the world. Her presence in the Department of Embryology broadened the scope of departmental research by acquainting us with the importance of plant biology and genetics. In 1994, she accepted the leadership role in a new biotechnology institute at Pennsylvania State University.

Steven McKnight came to the Department in 1977 as a postdoctoral fellow to work with Ron Reeder. When Reeder left in 1980, McKnight was made an independent staff associate. Richard Axel and his colleagues at Columbia had demonstrated how to introduce cloned genes into cultured animal cells. In those early days of recombinant DNA research, animal viruses were important model systems. The *Herpes simplex* virus in particular has a gene called thymidine kinase that confers resistance to a pharmacological agent in cells. McKnight developed the resistance assay and then carried out a systematic analysis of the region in the viral DNA that controls the expression of the gene.[26] With the aid of a technician and one undergraduate student, McKnight completely scooped several other very large labs that were trying to do the same experiment. We could not offer him a staff position, so with great reluctance we saw him hired by the Fred Hutchinson Cancer Research Institute in Seattle. However, he never left our radar screen, and in 1984 we hired him back as a staff member. By that time he no longer was studying the DNA regions that control genes but was studying the proteins in cells that interact with these same DNA elements. A few years after his return, McKnight and his colleagues discovered a new family of transcription factors that he named "leucine zipper" proteins.[27] These proteins bind as dimers to

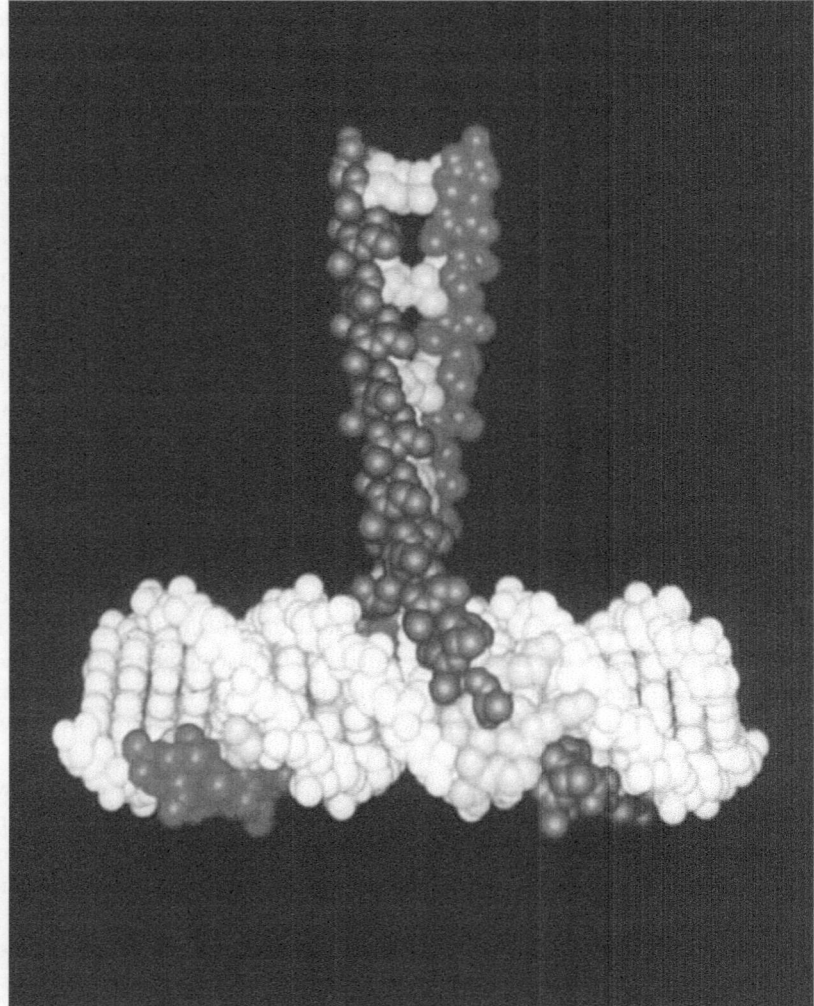

Figure 7.7 This model of a leucine zipper transcription factor was proposed by Steve McKnight and his colleagues. The protein functions as a dimer with the two chains held together by spaced leucine amino acids. The other ends of the protein molecules bind to a specific DNA sequence and influence the expression of a nearby gene.

a specific DNA controlling element. McKnight worked out the structure of these proteins and how equally spaced leucines in these proteins can function as a zipper to bind two proteins together into a functional unit (Fig. 7.7). In 1992, McKnight and two colleagues from California founded a biotech company called Tularik for the express purpose of developing drugs that interfere

Figure 7.8   Gerry Rubin (left) and Allan Spradling (right) in 1980, discussing how P-element mediated transformation works.

with the growth of cells by inhibiting gene expression. McKnight became the Director of Research at the company. Recently, he has returned to academia as the Chairman of the Department of Biochemistry at the University of Texas, Southwestern Medical School, in Dallas.

In 1980, two young geneticists, Gerald Rubin and Allan Spradling, were hired as staff members (Fig. 7.8). Spradling was trained in cell biology at the Massachusetts Institute of Technology and carried out a postdoctoral fellowship in *Drosophila* development and genetics at Indiana University with Anthony Mahowald. As a postdoctoral fellow, Spradling discovered the first example of amplification of genes that encode proteins rather than ribosomal RNAs. This genetic mechanism contributes to the high rates of protein synthesis by the follicle cells that surround the developing oocyte. At CIW, Spradling found that these amplifying genes made a good model for understanding chromosome replication.[28] Rubin generated one of the first sequences of a yeast gene during his graduate studies at the MRC labs. As a postdoctoral fellow at Stanford with Dave Hogness, he was among the first to address problems of *Drosophila* genetics using recombinant DNA technology. Rubin joined the Department after spending a few years on the faculty at Harvard Medical School where he had initiated molecular studies of *Drosophila* transposons, work he continued in Baltimore.

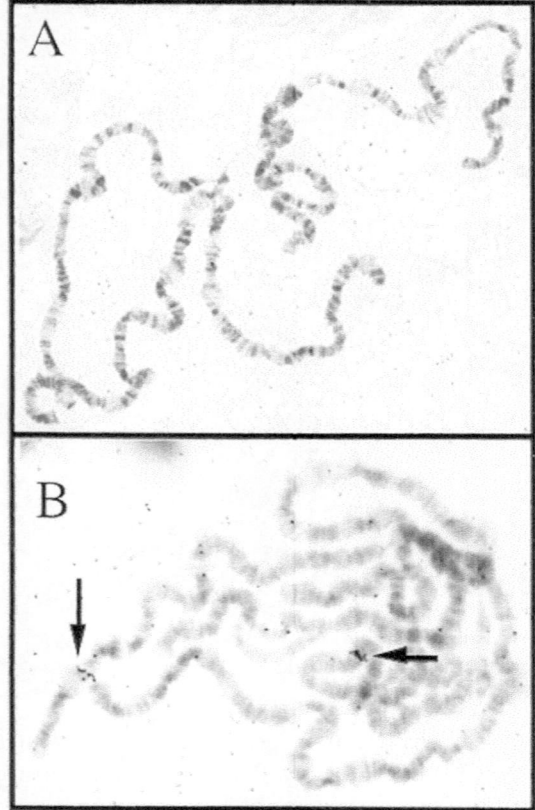

Figure 7.9 A picture of the large chromosomes of one of the original transgenic *Drosophila* prepared by the method of "P-element" transposition. DNA containing an intact P element had been injected into an embryo. A, Control chromosomes. B, The foreign DNA in these chromosomes is seen as black dots or grains (arrows).

Soon after their arrival, Spradling and Rubin asked me as Director to buy microinjection equipment so they could try a new project. They had devised a way to introduce genes into the eggs of *Drosophila*. The plan was to link a particular transposable element, like the one discovered by Barbara McClintock, to a gene of interest and then to inject the construct into the fertilized *Drosophila* egg. I doubt whether any investment by the Institution has ever had the immediate spectacular payoff as did our purchase of this equipment. Their method to transform *Drosophila* revolutionized *Drosophila* research[29] (Fig. 7.9). Because of this new tool, *Drosophila* became the organism used to identify the genes responsible for controlling early embryogenesis, discoveries that resulted in just the second Nobel Prize ever awarded for research in developmental biology to Eric Wieschaus and Christianne Nusslein-Volhard. Any

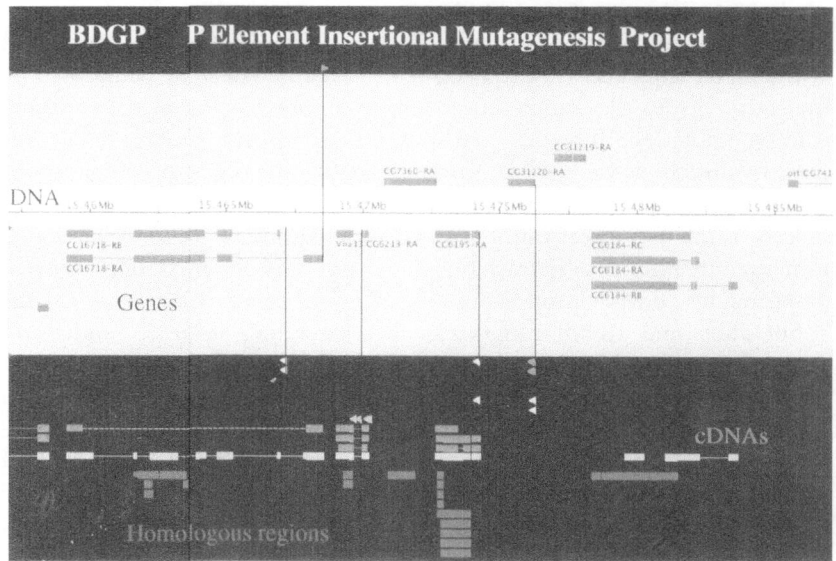

Figure 7.10 A sample of the computer interface used to identify P-element-induced mutations in *Drosophila* genes. This method was developed in the Spradling lab and then applied on a large scale by the Berkeley *Drosophila* Genome Project (BDGP). The black line in the center represents a small region of the *Drosophila* genome with genes shown as gray boxes above and below the line. Triangles connected to genes by vertical lines locate individual P elements. About 45% of all genes have been mutated to date by this method. The structure of cloned RNA transcripts (cDNAs) and regions homologous to genes in other organisms is shown in the lower part of the figure.

normal or mutant gene can be prepared by the recombinant DNA technology and then introduced stably into the fly genome. Even before they published their results in *Science*, Spradling and Rubin were teaching the community of *Drosophila* researchers how to microinject DNA into fertilized eggs and providing them with all of the tools. This generosity and openness has always been a hallmark of CIW scientists. Rubin stayed in the Department for just a few years. In 1983 he accepted an endowed professorship at the University of California, Berkeley.

In 1984, the Spradling lab began using the insertion of a new transposon as a method of mutagenesis that could at the same time mark the mutated gene for isolation. They discovered a way to eliminate the need for microinjection, so that many thousands of stocks containing a single transposon insertion at a new random location could be generated, simply by crossing flies (Fig. 7.10). Unlike strains with mutations induced by chemicals or radiation, the gene responsible for the mutant defects within any one of these "insertion" strains could be instantly located at the transposon insertion site.[30] A second approach to correlating genes and mutations involved mapping all the genes in the relevant chromosome region, ideally with the aid of the DNA sequence. An extra copy of each candidate gene could then be transformed into the mutant strain until the one was found that restored wild type function. Together, "transposon tagging" and "mutant rescue," as these two methods were called, made *Drosophila* the organism of choice for identifying the genes that control embryonic development. Inspired by this success in the fly, similar transposon tagging methods were soon developed in many other genetically tractable plants and animals, including a system pioneered by Nina Fedoroff for use in the model genetic plant *Arabidopsis*.

Spradling and Rubin recognized that extending these techniques to encompass the entire *Drosophila* genome would speed the application of genetic approaches to a myriad of biological questions. This realization inspired the Berkeley *Drosophila* genome project in 1993. Since then, the project has generated and distributed publicly many of the standard tools of modern *Drosophila* genetics, including the genomic DNA sequence, gene structures derived from transcript sequencing, and transposon insertion strains that disrupt nearly half of the approximately 13,500 *Drosophila* genes. Insertion strains aimed at the remaining genes continue to be generated in the Department. All these efforts have been enhanced greatly by a constant influx of information and resources from the *Drosophila* research community, who have shared their own research results, and developed many improvements to these methods.

In addition to their many contributions that benefit the entire *Drosophila* community, both Rubin and Spradling maintain active laboratories that probe important developmental questions. At Berkeley, Rubin has studied how the eye is formed. In the process, he demystified developmental inductions by showing that they result from widely studied intercellular signaling pathways that are commonly used for many purposes throughout organismal life. Spradling's group has continued to study egg development. They advanced the knowledge of gene amplification by identifying specific control sequences, replication origins, and regulatory genes.

More recently, interest in the Spradling lab has turned to early steps in egg development. The simple anatomy and powerful available genetic tools have allowed fundamental questions about stem cell regulation to be answered that are difficult to address in mammals. Another area of major focus has been the formation of oocytes within groups of interconnected germ cells known as germ line cysts. Many important components of *Drosophila* eggs originate in the other cyst cells. A fraction of their organelles and gene transcripts is transported into the future oocyte under the control of a special, polarized cytoskeletal system known as the fusome. The reasons for this novel process are still unproven, but it appears to be conserved among a wide range of organisms. Spradling has postulated that the transport process is

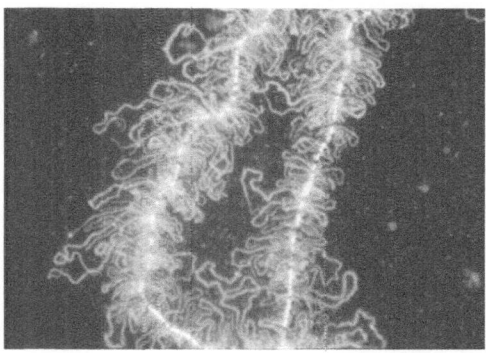

Figure 7.11 A portion of a giant lampbrush chromosome of the newt *Notophthalmus* stained pink with an antibody against a loop protein. The DNA axis appears in white. The unusually large size of these chromosomes and other nuclear structures in amphibian oocytes makes them ideal for the study of nuclear function and structure.

selective and facilitates the assembly of egg cytoplasm that is relatively free of defects.

In 1981, Joseph Gall, one of the world's most famous cytogeneticists, spent a sabbatical year in the Department. Upon returning to Yale he faced the ominous realization that it was his turn to become Chairman of their large Department of Biology. Gall was a superb teacher but above all he loved working in the lab. This was becoming increasingly difficult with his many university responsibilities and his large research group to supervise. In 1983, we sought a replacement for Douglas Fambrough and realized that Gall was an ideal "Carnegie" scientist. He accepted our offer immediately. This meant a reduction in the size of his research group but a return to unimpeded bench science. For years, Gall and his colleague in Scotland, H. G. Callan, had studied the giant chromosomes that are present uniquely in the nuclei of amphibian oocytes (Fig. 7.11). These developing egg cells contain enlarged nuclei that have the same structures and metabolic activities as normal-sized nuclei, making them excellent candidates for studying nuclear function. In the 1960s at Yale, Gall had been a co-discoverer of gene amplification, and he and a graduate student had developed the important method for assessing gene expression in cells at the cytological level. This technique, *in situ* hybridization, is used today extensively with little change. (For example, it is the method used in the experiments shown in Figs. 7.9 and 7.12.) With a few dedicated colleagues, Gall has continued his research on the contents and metabolism of oocyte nuclei. He has recently characterized an abundant nuclear structure first described almost a century ago by Ramon y Cajal. These "Cajal bodies" have all of the components for gene expression and may be a staging place for the proteins that transcribe genes.[31]

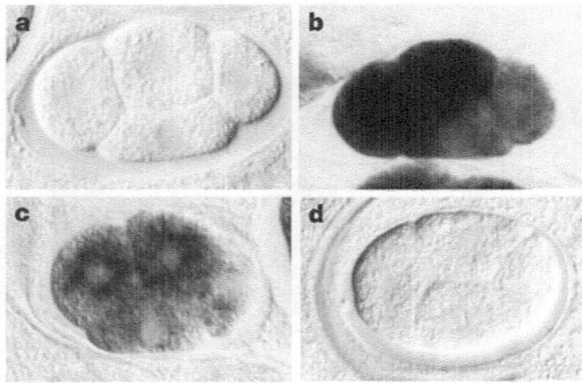

Figure 7.12 One of the first examples of specific RNA interference (RNAi). (a) A normal embryo that has not been reacted with the gene probe. (b) The RNA produced by a gene called *mex-3* in an early *Caenorhabditis elegans* embryo reacts with the gene probe in a control embryo and turns black. (c) An embryo injected with antisense RNA has a reduced signal, but (d) the embryo exposed to RNAi has no signal.

By the end of the 1980s, the small round worm, *Caenorhabditis elegans*, was an important model organism for developmental biology. Our resident worm expert, Sam Ward, had departed, and the search for Ward's replacement looked closely at genetically oriented applicants who studied this organism. Andrew Fire had come to the Department as a staff associate in 1986. Fire received his Ph.D. from MIT and, like Sam Ward, had carried out a post-doctoral period at the MRC labs in Cambridge. During that time he had developed a novel method to introduce genes into the syncytial gonad of the worm.[32] This discovery greatly expanded the usefulness of *C. elegans* for modern genetics, just as Spradling and Rubin's discovery of the use of transposable elements had done for *Drosophila*. As we compared applicants from around the world, we realized that we already had the best candidate in our Department. In 1989, Fire was appointed as a staff member. In the last ten years, Fire and his colleagues have made some of the most important discoveries on the control of gene expression. One of Fire's plans has been to identify the entire genetic pathway that leads to muscle formation in the worm. Along the way to this goal he discovered a general method for interfering with gene function. The administration of double stranded RNA homologous to a stretch of a given messenger RNA, called RNA interference or RNAi, can target the product of any gene for inactivation (Fig. 7.12).[33] Normally, RNA interference machinery appears to be deployed as a defense mechanism. This discovery is of an entirely novel mechanism for the control of gene expression. While the full significance of RNAi *in vivo* is just beginning to emerge, it has already revolutionized genetic studies. The method is so reliable that it is being used currently to screen all of the genes in the worm systematically for function. The genes that encode the RNAi machinery are found throughout much of the plant and animal kingdoms, and the method for suppressing gene function is already known to work successfully in many invertebrates and vertebrates RNAi methodology brings the power of genetic analysis to cultured cells, undoubtedly the biggest advance in tissue culture since its discovery. Perhaps most remarkable of all, this method permits investigators who study genes in organisms for which no traditional genetics exists to inhibit specifically gene function. The advent of such "instant genetics" may end the reign of a few model organisms and allow the full diversity of life to be studied on the frontiers of research. Without a doubt, the use of RNAi to inhibit gene function is the single most important method in biology that has been discovered in the last decade.

Douglas Koshland received his Ph.D. from MIT and carried out post-doctoral studies with Lee Hartwell in the Department of Genetics at the University of Washington. He became a staff member in 1987. Koshland recognized that many of the most important physiological events in the cells of higher organisms also take place in the relatively simple single cell yeast *Saccharomyces cerevisciae*. During mitosis, replicated chromosomes are seg-regated to opposite poles of the cell to ensure that each new daughter cell gets a complete copy of all the chromosomes. In preparation for this segre-gation event, chromosomes are modified, imparting to them a higher-order organization. This organization includes condensation, a specialized com-paction that facilitates chromosome movement, and cohesion between each pair of replicated chromosomes, which is essential for their correct sorting to opposite poles. While condensation and cohesion had been observed over 100 years ago, the molecular basis remained unknown largely because of lack of knowledge of the machinery (specific proteins) necessary for these pro-cesses. Koshland's lab has pioneered the identification and characterization of the proteins that mediate cohesion and condensation.[34] These discoveries provide one of the first clues that will form the basis for molecular under-standing of higher-order chromosome organization. The presence on the staff of the Department of Embryology of a geneticist–cell biologist studying an organism that does not even undergo development provides a hint to the evolving philosophy and reality of the Department and the research enter-prise in general. Whether they study yeast, plants, fruit flies, or the mouse, modern scientists share the methods and the important questions of biology. Scientists have learned to select the organism best suited for their particular inquiry. A department with the diversity of ours can and does have a common language. This mélange of research interests has made the Department an even more attractive place for independent scientists.

In 1993, the year before I gave up the directorship, five of the eight staff members were members of the US National Academy of Sciences (Gall,

Fedoroff, Spradling, McKnight, and myself). A number of us had won awards for our research. The challenge of the new Director, Allan Spradling, would be to maintain the high quality of the Department in the face of a rapidly changing research enterprise.

Some of the greatest new opportunities are likely to be found in the neuro-sciences. The brain, that most complex of tissues, is still poorly characterized even at the cellular level. These frontiers have attracted many outstanding young scientists to the study of brain development in recent years. One such person, Marnie Halpern, was appointed to our staff in 1994. Following her Ph.D. from Yale studying *Drosophila* development, Halpern changed her research interest to the zebrafish, a newly established model organism for developmental biologists. As a postdoctoral fellow at the University of Oregon, Halpern characterized early-acting mutations that affect tissue pat-terning of the embryo. The zebrafish embryo is almost completely trans-parent and embryonic development occurs in about three days. These and other features have elevated zebrafish as a model vertebrate for research in embryology.

Halpern's initial research clarified the relationship of the notochord to structures in the developing nervous system. Zebrafish mutants that lack a notochord still form the floor plate and have a relatively normal nervous system. In contrast, a specialized class of muscle cells, the muscle pioneers, is dependent on the presence of a notochord. Her research involves the signal pathways that have been found by modern genetic biologists to account for the interactions of cells and tissues in development. These cell–cell inter-actions explain many of the phenomena that had been discovered decades earlier by the experimental embryologists. One of these, the "nodal" signal pathway, influences the asymmetric placement of visceral organs like the heart and intestinal tract. Recently, Halpern has implicated components of this pathway in asymmetric features of the zebrafish brain.[35] An exciting dis-covery is that an asymmetry in one region can influence the gene expression in a neighboring brain nucleus, thus propagating left–right differences in the brain. Whether similar mechanisms are responsible for the anatomical and perhaps even the functional differences of the left and right sides of the human brain is a question that can now be pursued using the molecular tools developed in the zebrafish.

One of the most important model organisms used increasingly in modern developmental biology has been the mouse. There is a long history of mouse genetics, and the mouse is the experimental animal most closely related to humans. However, in 1994, the Department lacked a modern facility for mouse research, including the labs and equipment needed to generate transgenic and knockout mice. The lack of appropriate facilities did not deter Susan Dymecki, who had been appointed as a staff associate in 1993. She housed her mice in the Department's old incinerator. Within a few years she developed a novel way to rearrange genes under experimental control in living mice. Dymecki has continued to perfect this technology and apply it to the study of brain development since joining the faculty of Harvard Medical School in 1996. Upon becoming Director in 1994, Allan Spradling obtained a commitment from the Institution to build a modern state-of-the-art facility to house mice for genetic research. With Sue Dymecki's aid it was completed in 1995.

Our first staff member who specializes in mouse embryology, Chen-Ming Fan, was appointed in 1995. He had carried out his graduate research at Harvard followed by a postdoctoral period learning how to work with mouse embryos at the University of California at San Francisco. The basic pattern of the mammalian muscle and skeletal system is laid out during embryonic development. All skeletal muscles and much of the bone originate from a group of precursor cells called the somites. The patterning of bone, muscle, and dermis (the layer below the skin) is dependent entirely on signals that come from adjacent tissues. This fact was established in the days of experimental embryology, but a challenge of the modern era has been to identify the molecular nature of these signals and the underlying mechanism of their action. Fan has identified many of the genes involved in these inductive reactions.[36] Their ability to orchestrate the development of various somitic components in a correct pattern relies on a delicate balance of their concerted actions. Such a precise balance is accomplished through a complex succession of mutually antagonistic, as well as synergistic, activities.

One advantage of the mouse system is the ability to test gene function in a highly precise manner. Genes can be mutated to virtually any desired form or completely "knocked out" by deletion at their normal site on the chromosome. First, the mutant form of the gene is constructed *in vitro* and introduced into embryonic stem cells. These stem cells are then used to create mutant mice lacking the function of these genes. Fan has used this remarkable combination of molecular genetics and cell biological methods to investigate the role played by two closely related transcription factors during midbrain development. Although both are related to a *Drosophila* midbrain gene, his research has documented distinct roles for the two mouse orthologs.

In 1996, Yixian Zheng, a young cell biologist who had just completed her postdoctoral research at the University of California San Francisco Medical School, was appointed as a staff member. Her research on the crucial question of how chromosomes are separated with each cell division has been at the forefront of this important subject. A spindle apparatus attaches to chromosomes and pulls them apart. The spindle consists of several components including a microtubule cytoskeleton that is connected to structures called centrosomes. The duplication of the centrosome at each cell cycle ensures formation of a bipolar spindle and inheritance of a centrosome by each daughter cell. Mistakes in any of these processes will lead to cell cycle arrest. The resulting mis-segregation of chromosomes can be the basis of birth defects. Zheng and her colleagues have been identifying systematically the molecules that make up these indispensable cell structures and how they interact to ensure that each daughter cell receives the exact number of intact chromosomes. It has been difficult to elucidate the mechanism of microtubule nucleation and spindle assembly because it involves many structural and regulatory components. To overcome the complexity of studying microtubule nucleation by centrosomes, Zheng and her colleagues have purified a key component called γ-Tubulin Ring Complex (γ-TuRC) from *Xenopus* egg and *Drosophila* embryo extracts and have developed assays to show that γ-TuRC is the long sought after microtubule nucleator.[37] In this continuing detective story, the Zheng laboratory has discovered several other crucial gene products that are essential for spindle assembly at mitosis.

The staff associates

As I mentioned earlier, it has long been departmental tradition to provide space and facilities for independent scientists to spend time in the Department before either returning to a preexisting job or moving to a tenure track appointment. Both Igor Dawid and I had come to the Department as independent "fellows" with no specified mentor. After becoming Director, I decided to formalize these independent junior faculty positions. "Staff associates" are promising young scientists either directly out of graduate school or who have already had a postdoctoral period. The staff associate position was defined as entirely independent with no ties to any staff members and from three to five years in duration. A staff associate is expected to work full time in the lab with the assistance of a technician. Staff associates are discouraged from having graduate students or postdoctoral fellows. We expect that once a staff associate's research is up and running he/she will then search for a tenure track position elsewhere with the opportunity for students. A staff associate is not on track to become a staff member. They can enter a competition for a staff member vacancy, but they have no advantage over other applicants in a worldwide search. Two staff associate modules were built as part of a second floor renovation project in 1987, and two additional ones were added in a second remodeling in 1995. The success of these independent young scientists has certainly surpassed our wildest dreams (Table 7.1). Having several energetic young scientists with different research programs at any given time invigorates the Department. The staff associate program has now been copied in several other research institutions and universities.

A summary

For almost ninety years the Department of Embryology has been a secret jewel amongst Baltimore institutions. Few non-scientists in Baltimore know

Table 7.1 *Staff associates of the Department of Embryology 1976–2002*

Name	Years	Current Title/Position
Pat Gearhart	1979–81	National Inst. of Aging, National Inst. of Health
Steve McKnight	1980–2	Chairman, Biochemistry Department, University of Texas Southwestern Medical Center
Rick Rotundo	1982–4	Professor, Department Cell Biology & Anatomy, University of Miami
Sondra Lazarowitz	1983–90	Professor, Department of Plant Pathology, Cornell University
Martin Snider	1983–5	Prof., Biochemistry, Case Western Reserve
David Schwartz	1985–9	Chemistry Dept., NYU
Phil Beachy	1986–8	Prof., Mol. Biology, JHU School of Medicine & Investigator, Howard Hughes Medical Institute
Andy Fire	1986–9	Staff Member, Dept. of Embryology, Carnegie Inst. of Washington
Se-Jin Lee	1989–91	Professor, Molecular Biology & Genetics, JHU School of Medicine
Denise Montell	1989–92	Professor, Dept. of Biological Chemistry, JHU School of Medicine
Nipam Patel	1991–5	Associate Investigator, Howard Hughes Medical Inst., The University of Chicago
Catherine Thompson	1992–6	Asst. Prof., Department of Neuroscience, JHU School of Medicine/Kennedy Krieger Institute
Susan Dymecki	1993–8	Asst. Prof., Department of Genetics Harvard Medical School
Pernille Rorth	1995–8	Res. Scientist, Developmental Biology, EMBL-Heidelberg Program
Alejandro Sanchez	1996–2002	Asst. Prof., Dept. of Neurobiology & Anatomy, Univ. of Utah School of Medicine
Jimo Borjigin	1998–present	CIW, Embryology
Erika Matunis	1998–2002	Asst. Prof. Dept. of Cell Biology JHU School of Medicine
Terence Murphy	1999–present	CIW, Embryology
Jim Wilhem	2001–present	CIW, Embryology

NYU, New York University; JHU, The Johns Hopkins University; EMBL, European Molecular Biology Laboratory.

of our existence. Our identification with Washington helps to maintain our anonymity. In 1973, I published an article in *Scientific American*. It listed my affiliation as the Carnegie Institution of Washington. The vast majority of the more than a thousand reprint requests were sent to Washington. Visitors to our Department arriving by taxi from airports or the train station are advised to instruct the taxi driver with our address rather than our name. This invisibility has guaranteed that we will be much better known around the world than in Baltimore. A compliment that we receive from first time scientific visitors has been "I thought that this Department was much larger than it is."

Over the years the animals that have been studied have reflected the evolution in the Department's research. In 1960, the experimental animals were the chicken, guinea pig, rabbit, monkey, and opossum. In 1976, the animals were frogs, cultured cells, chickens, and leeches. In the 1990s, the organisms studied in the Department were frogs, plants, *Drosophila*, *C. elegans*, yeast, mouse, and zebrafish. This progression inexorably to lowlier model organisms coincides with the rise of genetics as the most important discipline for understanding developmental phenomena. The diversity of organisms studied in the Department is not an accident. Not only does each investigator study a different system, but also this variety represents a strong faith in the success of individuals working with just a few colleagues. In the second half of the twentieth century the Department has fulfilled the future anticipated for it by George Corner and the confidence placed in it by the Institution that a small department could evolve and play a meaningful role in its field. More than that it has led the remarkable evolution of the field of embryology from an anatomic specialty in 1950 into the main stream of modern biology in 2000.

Notes

1. G. Corner, Carnegie Institution of Washington *Year Book* 54 (1954–5), p. 192.
2. A. M. Mun, J. Errico, and J. D. Ebert, "Ontogeny of the graft-versus-host reaction in the chick: immunological maturation of spleen cells perpetuated, by serial transfer, in an embryonic environment," *The Anatomical Record* 139 (1961), p. 258.
3. D. W. Bishop and H. Hoffmann-Berling, "Extracted mammalian sperm models, 1. Preparation and reactivation with adenosine triphosphate," *Journal of Cellular and Comparative Physiology* 53 (1954), pp. 445–66.
4. B. G. Boving, "Implantation," *Annals of the New York Academy of Science* 75 (1959), pp. 700–25.
5. R. K. Burns and L. M. Burns, "Observations on the breeding of the American opossum in Florida," *Revue Suisse de Zoology* 64 (1957), pp. 595–605.
6. R. L. DeHaan, "Time-lapse photographic analysis of migration of the pre-cardiac mesoderm in the early chick embryo," *American Zoologist* (1961), pp. 444–5.

7. M. E. Rawles, "Tissue interactions in scale and feather development as studied in dermal–epidermal recombinations," *Journal of Embryology and Experimental Morphology* 11 (1963), pp. 765–89.

8. D. D. Brown and J. B. Gurdon, "Absence of ribosomal RNA synthesis in the anucleolate mutant of *Xenopus laevis*," *Proceedings of the National Academy of Science* 51 (1964), pp. 139–46.

9. H. Wallace and M. Birnstiel, "Ribosomal cistrons and the nucleolar organizer," *Biochimica Biophysica Acta* 114 (1966), pp. 296–310.

10. I. B. Dawid, "Deoxyribonucleic acid in amphibian eggs," *Journal of Molecular Biology* 12 (1965), pp. 581–99.

11. D. D. Brown and I. B. Dawid, "Specific gene amplification in oocytes," *Science* 160 (1968), pp. 272–80.

12. J. G. Gall, "Differential synthesis of the genes for ribosomal RNA during amphibian oogenesis," *Proceedings of the National Academy of Science* 60 (1968), pp. 553–60.

13. J. F. Morrow, S. N. Cohen, A. C. Y. Chang, H. Boyer, H. M. Goodman, and R. B. Helling, "Replication and transcription of eukaryotic DNA in *Escherichia coli*," *Proceedings of the National Academy of Science* 71 (1974), pp. 1743–7.

14. R. H. Reeder, D. D. Brown, P. K. Wellauer, and I. B. Dawid, "Patterns of ribosomal DNA spacer lengths are inherited," *Journal of Molecular Biology* 105 (1976), pp. 507–16.

15. P. K. Wellauer and I. B. Dawid, "Secondary structure maps of ribosomal RNA and DNA I," *Journal of Molecular Biology* 89 (1974), pp. 379–95.

16. Y. Suzuki and D. D. Brown, "Isolation and identification of the messenger RNA for silk fibroin from *Bombyx mori*," *Journal of Molecular Biology* 63 (1972), pp. 409–29.

17. D. D. Brown, P. C. Wensink, and E. Jordan, "Purification and some characteristics of 5S DNA from *Xenopus laevis*," *Proceedings of the National Academy of Science* 68 (1971), pp. 3,175–9.

18. S. Sakonju, D. F. Bogenhagen, and D. D. Brown, "A control region in the center of the 5S RNA gene directs specific initiation of transcription: I. The 5′ border of the region," *Cell* 19 (1980), pp. 13–25.

19. A. Wolffe and D. D. Brown, "Developmental regulation of two 5S ribosomal RNA genes," *Science* 241 (1988), pp. 1,626–32.

20. I. R. Konigsberg, "Clonal analysis of myogenesis," *Science* 140 (1963), pp. 1,273–84.

21. D. M. Fambrough, D. B. Drachman, and S. Satyamurti, "Neuromuscular junction in myasthenia gravis: decreased acetylcholine receptors," *Science* 182 (1973), pp. 293–5.

22. R. E. Pagano, K. Longmuir, O. C. Martin, and D. K. Struck, "Metabolism and intracellular localization of a fluorescently labeled intermediate in lipid biosynthesis within cultured fibroblasts," *Journal of Cell Biology* 91 (1981), pp. 872–7.

23. K. J. Muller and U. J. McMahan, "The shape of sensory and motor neurons and the distribution of their synapses in ganglia of the leech: a study using intracellular injection of horseradish peroxidase," *Proceedings of the Royal Society of London, Ser. B* 194 (1976), pp. 481–99.

24. D. J. Burke and S. Ward, "Identification of a large multigene family encoding the major sperm protein of *Caenorhabditis elegans*," *Journal of Molecular Biology* 171 (1983), pp. 1–29.

25. N. Fedoroff, S. Wessler, and M. Shure, "Isolation of the transposable maize controlling elements *Ac* and *Ds*," *Cell* 35 (1983), pp. 243–51.

26. S. L. McKnight and E. R. Gavis, "Expression of the *Herpes simplex* virus thymidine kinase in *Xenopus laevis* oocytes: an assay for the study of deletion mutants constructed in vitro," *Nucleic Acids Research* 8 (1980), pp. 5,931–48.

27. W. H. Landschulz, P. F. Johnson, and S. L. McKnight, "The leucine zipper: a hypothetical structure to a new class of DNA binding proteins," *Science* 240 (1988), pp. 1,759–64.

28. A. C. Spradling, "The organization and amplification of two chromosomal domains containing *Drosophila* chorion genes," *Cell* 27 (1981), pp. 193–201.

29. A. C. Spradling and G. M. Rubin, "Transposition of cloned P elements into *Drosophila* germ line chromosomes," *Science* 218 (1982), pp. 341–7.

30. L. Cooley, R. Kelley, and A. C. Spradling, "Insertional mutagenesis of the *Drosophila* genome with single P elements," *Science* 239 (1988), pp. 1,121–8.

31. J. G. Gall, M. Bellini, Z. Wu, and C. Murphy, "Assembly of the nuclear transcription and processing machinery: Cajal bodies (coiled bodies) and transcriptosomes," *Molecular Biology of the Cell* 10 (1999), pp. 4,385–402.

32. A. Fire, "Integrative transformation of *Caenorhabditis elegans*," *The EMBO Journal* 5 (1986), pp. 2,673–80.

33. A. Fire, S. Xu, M. K. Montgomery, S. A. Kostas, S. E. Driver, and C. C. Mello, "Potent and specific genetic interference by double stranded RNA in *C. elegans*," *Nature* 391 (1998), pp. 806–11.

34. B. D. Lavoie, E. Hogan, and D. Koshland, "*In vivo* dissection of the chromosome condensation machinery: reversibility of condensation distinguishes contributions of condensin and cohesin," *Journal of Cell Biology* 156 (2002), pp. 805–15.

35. J. O. Liang, A. Etheridge, L. Hantsoo, A. L. Rubinstein, S. J. Nowak, J. C. Izpisua Belmonte, and M. E. Halpern, "Asymmetric nodal signaling in the zebrafish diencephalon positions the pineal organ," *Development* 127 (2000), pp. 5,101–12.

36. C. S. Lee, L. Buttitta, and C. M. Fan, "Evidence that the WNT-inducible growth arrest-specific gene 1 encodes an antagonist of sonic hedgehog signaling in the somite," *Proceedings of the National Academy of Science* 98 (2001), pp. 11,347–52.

37. A. Wilde, S. B. Lizarraga, L. Zhang, C. Wiese, N. R. Gliksman, C. E. Walczak, and Y. Zheng, "Ran stimulates spindle assembly by altering microtubule dynamics and the balance of motor activities," *Nature Cell Biology* 3 (2001), pp. 221–7.

Bibliography

Bishop, D. W. and H. Hoffmann-Berling, "Extracted mammalian sperm models, 1. Preparation and reactivation with adenosine triphosphate," *Journal of Cellular and Comparative Physiology* 53 (1954), pp. 445–66.

Boving, B. G., "Implantation," *Annals of the New York Academy of Science* 75 (1959), pp. 700–25.

Brown, D. D. and I. B. Dawid, "Specific gene amplification in oocytes," *Science* 160 (1968), pp. 272–80.

Brown, D. D. and J. B. Gurdon, "Absence of ribosomal RNA synthesis in the anucleolate mutant of *Xenopus laevis*," *Proceedings of the National Academy of Science* 51 (1964), pp. 139–46.

Brown, D. D., P. C. Wensink, and E. Jordan, "Purification and some characteristics of 5S DNA from *Xenopus laevis*," *Proceedings of the National Academy of Science* 68 (1971), pp. 3,175–9.

Burke, D. J. and S. Ward, "Identification of a large multigene family encoding the major sperm protein of *Caenorhabditis elegans*," *Journal of Molecular Biology* 171 (1983), pp. 1–29.

Burns, R. K. and L. M. Burns, "Observations on the breeding of the American opossum in Florida," *Revue de Suisse Zoology* 64 (1957), pp. 595–605.

Cooley, L., R. Kelley, and A. C. Spradling, "Insertional mutagenesis of the *Drosophila* genome with single P elements," *Science* 239 (1988), pp. 1,121–8.

Corner, G., Carnegie Institution of Washington *Year Book* 54 (1954–5), p. 192.

Dawid, I. B., "Deoxyribonucleic acid in amphibian eggs," *Journal of Molecular Biology* 12 (1965), pp. 581–99.

DeHaan, R. L., "Time-lapse photographic analysis of migration of the pre-cardiac mesoderm in the early chick embryo," *American Zoologist* (1961), pp. 444–5.

Fambrough, D. M., D. B. Drachman, and S. Satyamurti, "Neuromuscular junction in myasthenia gravis: decreased acetylcholine receptors," *Science* 182 (1973), pp. 293–5.

Fedoroff, N., S. Wessler, and M. Shure, "Isolation of the transposable maize controlling elements *Ac* and *Ds*," *Cell* 35 (1983), pp. 243–51.

Fire, A., "Integrative transformation of *Caenorhabditis elegans*," *The EMBO Journal* 5 (1986), pp. 2,673–80.

Fire, A., S. Xu, M. K. Montgomery, S. A. Kostas, S. E. Driver, and C. C. Mello, "Potent and specific genetic interference by double stranded RNA in *C. elegans*," *Nature* 391 (1998), pp. 806–11.

Gall, J. G., "Differential synthesis of the genes for ribosomal RNA during amphibian oogenesis," *Proceedings of the National Academy of Science* 60 (1968), pp. 553–60.

Gall, J. G., M. Bellini, Z. Wu, and C. Murphy, "Assembly of the nuclear transcription and processing machinery: Cajal bodies (coiled bodies) and transcriptosomes," *Molecular Biology of the Cell* 10 (1999), pp. 4,385–402.

Konigsberg, I. R., "Clonal analysis of myogenesis," *Science* 140 (1963), pp. 1,273–84.

Landschulz, W. H., P. F. Johnson, and S. L. McKnight, "The leucine zipper: a hypothetical structure to a new class of DNA binding proteins," *Science* 240 (1988), pp. 1,759–64.

Lavoie, B. D., E. Hogan, and D. Koshland, "*In vivo* dissection of the chromosome condensation machinery: reversibility of condensation distinguishes contributions of condensin and cohesin," *Journal of Cell Biology* 156 (2002), pp. 805–15.

Lee, C. S., L. Buttitta, and C. M. Fan, "Evidence that the WNT-inducible growth arrest-specific gene 1 encodes an antagonist of sonic hedgehog signaling in the somite," *Proceedings of the National Academy of Science* 98 (2001), pp. 11,347–52.

Liang, J. O., A. Etheridge, L. Hantsoo, A. L. Rubinstein, S. J. Nowak, J. C. Izpisua Belmonte, and M. E. Halpern, "Asymmetric nodal signaling in the zebrafish diencephalon positions the pineal organ," *Development* 127 (2000), pp. 5,101–12.

McKnight, S. L. and E. R. Gavis, "Expression of the *Herpes simplex* virus thymidine kinase in *Xenopus laevis* oocytes: an assay for the study of deletion mutants constructed in vitro," *Nucleic Acids Research* 8 (1980), pp. 5,931–48.

Morrow, J. F., S. N. Cohen, A. C. Y. Chang, H. Boyer, H. M. Goodman, and R. B. Helling "Replication and transcription of eukaryotic DNA in *Escherichia coli*," *Proceedings of the National Academy of Science* 71 (1974), pp. 1,743–7.

Muller, K. J. and U. J. McMahan, "The shape of sensory and motor neurons and the distribution of their synapses in ganglia of the leech: a study using intracellular injection of horseradish peroxidase," *Proceedings of the Royal Society of London, Ser.* B 194 (1976), pp. 481–99.

Mun, A. M., J. Errico, and J. D. Ebert, "Ontogeny of the graft-versus-host reaction in the chick: immunological maturation of spleen cells perpetuated, by serial transfer, in an embryonic environment," *The Anatomical Record* 139 (1961), p. 258.

Pagano, R. E., K. Longmuir, O. C. Martin, and D. K. Struck, "Metabolism and intra-cellular localization of a fluorescently labeled intermediate in lipid biosynthesis within cultured fibroblasts," *Journal of Cell Biology* 91 (1981), pp. 872–7.

Rawles, M. E., "Tissue interactions in scale and feather development as studied in dermal–epidermal recombinations," *Journal of Embryology and Experimental Morphology* 11 (1963), pp. 765–89.

Reeder, R. H., D. D. Brown, P. K. Wellauer, and I. B. Dawid, "Patterns of ribosomal DNA spacer lengths are inherited," *Journal of Molecular Biology* 105 (1976), pp. 507–16.

Sakonju, S., D. F. Bogenhagen, and D. D. Brown, "A control region in the center of the 5S RNA gene directs specific initiation of transcription: I. The 5′ border of the region," *Cell* 19 (1980), pp. 13–25.

Spradling, A. C., "The organization and amplification of two chromosomal domains containing *Drosophila* chorion genes," *Cell* 27 (1981), pp. 193–201.

Spradling A. C. and G. M. Rubin, "Transposition of cloned P elements into *Drosophila* germ line chromosomes," *Science* 218 (1982), pp. 341–7.

Suzuki, Y. and D. D. Brown, "Isolation and identification of the messenger RNA for silk fibroin from *Bombyx mori*," *Journal of Molecular Biology* 63 (1972), pp. 409–29.

Wallace, H. and M. Birnstiel, "Ribosomal cistrons and the nucleolar organizer," *Biochimica Biophysica Acta* 114 (1966), pp. 296–310.

Wellauer, P. K. and I. B. Dawid, "Secondary structure maps of ribosomal RNA and DNA I," *Journal of Molecular Biology* 89 (1974), pp. 379–95.

Wilde, A., S. B. Lizarraga, L. Zhang, C. Wiese, N. R. Gliksman, C. E. Walczak, and Y. Zheng, "Ran stimulates spindle assembly by altering microtubule dynamics and the balance of motor activities," *Nature Cell Biology* 3 (2001), pp. 221–7.

Wolffe, A. and D. D. Brown, "Developmental regulation of two 5S ribosomal RNA genes," *Science* 241 (1988), pp. 1,626–32.

8

LOOKING AHEAD

ALLAN SPRADLING

Department of Embryology, Carnegie Institution of Washington

A remarkable record

The primary concern of a Director at the Carnegie Institution of Washington (CIW) lies with the Department's future, not its past. Even after twenty-two years in the Department of Embryology, I had not known as much about its past as is revealed by the foregoing chapters. Most remarkable in retrospect is how successfully the Department has evolved from studying anatomical issues closely allied with human reproduction to studying fundamental biological questions at the cellular and molecular level. What possible relationship could a department that appeared to Don Brown in 1960 as "a traditional medical school anatomy department or perhaps a museum," have with the challenges of today's high tech, high visibility biological enterprise? In fact, certain elements do seem to have been held in common for the last ninety years by the scientists who thrived in the New Hunterian Building at Johns Hopkins Medical School and at 115 W. University Parkway. These shared features, largely reflective of the Institution as a whole, are paramount when contemplating why the Department of Embryology has thrived for so long and whether this tradition can serve us well as it heads into the new century and a new home on San Martin Drive (Fig. 8.1).

Few departments ever achieve the prominence enjoyed by the Department of Embryology and fewer still maintain such distinction for significant periods of time. For any scientific institution to produce world-class research over such a span of years requires above all that it be able to change. Establishing a new research direction often requires developing a new technique or novel instrument, and demands that one relentlessly pursue experiments with little foreseeable prospect of success in search of answers that may not even be important. Sadly, the patient, long-term support needed to launch such ventures is usually lacking even today in corporate R and D departments, in universities, in the US national granting system, and even in many

Figure 8.1   Artist's rendition of the Department of Embryology's new building.

research institutes, where the focus swings with the latest and most visible problems. The ability to nurture such innovation, and the will to follow it along a natural course of change, represents the key to long-term success.

One of the remarkable features of this Department is that it has managed to succeed using quite disparate styles and approaches to supporting research. In its early days, the Department excelled in a mission-oriented program to collect and analyze human embryos. Later, it prospered by developing and exploiting a large-scale, in-house research resource: its breeding monkey colony. In more recent times the Department has thrived without any special research program or unique technology. For much of its history the departmental decisions were made almost entirely by the Director, yet for the last thirty years it has functioned democratically. Until 1975, neither the Department (nor the Institution as a whole) accepted outside funds for fear that research independence might be compromised. Since then, beginning with NIH grants submitted by Don Brown and Igor Dawid, an increasing fraction of funds in the Department and in the CIW have come from sources other than the endowment. Yet even today, when two thirds of the departmental budget is derived from outside sources, there has been virtually no effect on the originality or productivity of the staff. What has the Department held to amidst so much change?

One factor that has not changed is a belief in the central importance of the individual staff member as an independent researcher. This Department has thrived consistently by finding and supporting exceptional individuals who yearn to venture off the edges of the known, with minimal regard for the time and resources that will be required. We are not successful because we can anticipate better than others what will be found there. Rather our philosophy is to recognize individuals with a glint in their eyes, to trust in their judgment, and to not listen too closely to the details. It is a continuing surprise that well-prepared, provisioned, and patiently supported individuals succeed more frequently than anyone can reasonably expect. Seemingly insurmountable barriers are regularly overcome by human ingenuity, and the answers to well-posed questions almost routinely turn out to be of much greater practical significance than predicted.

Thus, the Department of Embryology has been based on a shared set of values and an approach to research we call, for lack of a better term, "the Carnegie style." While widely admired in principle, in practice our Department embodies this style in perhaps its purest and most concentrated form. People who formerly worked here are its strongest advocates: "I did my best work while I was at Carnegie," "After leaving here I had to go into the real world," "You don't realize how lucky you are to work here," "If you can't do research here, you can't do it anywhere," "What my department just did would never happen at Carnegie," "They want to be like Carnegie, but they just don't understand."

One expression of the Carnegie style is how we utilize our research space. Each regular faculty member has only about 800 sq. ft. of personal laboratory space with a small, attached office. However, research groups larger than four or five house some of their members in common shared laboratories located nearby. Thus, for many, the researcher across the bench works in a different research group. For everyone, whether based in a main or shared lab, a great deal of research activity takes place in other common departmental space. These rooms contain all major research instruments. Facilities to house and care for research animals are also separate, though located strategically relative to their major users. In addition, the Department maintains several core facilities that are open and used by all. All departmental researchers help themselves to commonly used laboratory supplies and reagents that are maintained in common stock rooms.

One consequence of these arrangements is that researchers move about the building during the course of their daily routine. There is no sense of leaving and re-entering one's own domain traveling in and out of the home laboratory. Instead, one feels like a citizen of the entire building. Dispersed functions are more efficient since there is less duplication of resources than in many departments. Traveling to and working in common areas guarantees numerous informal contacts with other department members. This arrangement of workspaces and the overall small size ensure that no one is anonymous in our Department. A new face is spotted immediately.

The feeling that there are relatively few barriers to trying out a new idea is a critical component of the Carnegie style. New projects often require new instruments, new experimental animals, and new space to house these items. Importantly, faculty expect to be able to innovate much more quickly than in most other departments. Where researchers would have to write a grant and wait many months to initiate work following their decision to pursue a new direction, our faculty can and do expect to start immediately. The Department strives to retain enough flexible space and funds to accommodate unexpected projects. It is understood that any area that falls into disuse is likely to be renovated and assigned a new role.

Major rites of the Carnegie style take place each week at regular scientific meetings of the entire Department. In the most important meeting, one researcher each week presents the results of his/her recent investigations. These sessions are held in a relatively small room, and attendance is expected. Questions occur throughout the course of the presentation and may continue for a considerable time afterwards. A common sense of intellectual standards is developed at these meetings, and they enable one to recall a person's current accomplishments and problems at the time of their next encounter. Interesting conversations, ideas, and novel solutions are often born from this combination.

Scientific creativity is ultimately bolstered by our system of professional advancement. There is no tenure system. Instead, all faculty members are evaluated at five-year intervals, and the standard of these reviews is crucial. The originality and long-term significance of a research program is emphasized rather than its funding level or professional visibility. The lack of tenure ensures that a faculty member can never rest on past triumphs. This has a way of placing more pressure on older rather than younger faculty, which is the reverse of the situation in tenure track universities. Periodic review is a more forgiving system than tenure because mistaken decisions are not irreversible. Young investigators know that innovation is more cherished than quantity and the Department is patient in its support.

We are not impressed by the number of publications or the journals in which research is published. The recent Nobel prizes for *Drosophila* research recognized work that originally appeared in the *Roux Archives*, not in *Science* or *Nature*. I recently heard a distinguished scientist brag that each of his last three papers, published in the most prestigious journals, appeared almost simultaneously with virtually identical contributions from five other groups. He meant to validate the wide interest in his work and his great competitive skills. The Carnegie scientist, however, might note that scientific knowledge would have been little affected had he simply thrown his manuscripts into the trash.

Another important component of the Carnegie style is its democratic underpinnings. On the one hand, committees and unnecessary bureaucratic procedures are kept to an absolute minimum. However, mutual respect ensures that no major decision affecting the well-being of the Department will occur prior to consultation with the entire faculty. As a result, a strong department-wide consensus is usually achieved. The Director always retains the authority to settle matters when a democratic approach does not succeed, a situation that paradoxically appears to have a calming influence. Of course, the Carnegie style would be meaningless were it not practiced and supported by senior scientists and leaders throughout the Institution.

This research style is not for everyone and in fact some individuals do not accept it for a lifetime. One staff member observed that the Department is a bad place to retire from. Carnegie scientists must from time to time be willing to pull up their intellectual stakes and move on. The style is superb for starting new ventures but less successful at maintaining leadership once a field becomes well populated with other researchers and with questions susceptible to the straightforward application of existing techniques. It might be easier to get a big lab and a lot of grant support and remain relevant by sheer mass action, but it wouldn't be "Carnegie." If you are an early comer, it can be difficult to leave a field you helped start, where you are held in respect, and that may subsequently falter. But starting over turns one's attention back to the true frontiers and their possibilities. Remarkably, it is another example where high expectations are generally rewarded.

Prospects for the future

By the mid 1990s the Department had completed its evolution to studying the problems of development at the most fundamental levels using diverse systems. The faculty and department were widely recognized. A tradition of full participation and consensus among staff about important issues had taken hold and further enhanced our scientific lives. Consequently, the challenges facing the new Director were quite different from those addressed by the leaders who established the Department or led it in new directions. Excellence would have to be maintained in the midst of radical changes in the practice of biological research that were sweeping the field.

Many who have come to know our Department and its accomplishments in recent years express surprise that an institute with our organizational structure can consistently produce important and novel work. First, they suggest we are much too small. We lack a faculty member with expertise in each major biological discipline and technology. There are not enough faculty working on the same basic system to form a "critical mass." Yes, it may have worked in the past, but biology today has changed. Surely quality in such a small place represents an unstable equilibrium, soon to decay.

The critics are certainly right about one thing; biology has changed immensely during the last decade or two. The marriage of genetics and developmental biology that the Department helped foster has become the scientific mainstream. As it became possible to sequence entire genomes and most of their transcripts, hundreds of small biotech companies were formed. In the private sector a scientist can often obtain information on a gene of interest from various company departments in just days that would require weeks or months to acquire in an academic laboratory. Large collections of resources developed by these enterprises are usually not shared. "If you only knew how good the resources are here" a former Carnegie Staff Associate recalls being told recently by a smiling Vice President of Research at such a company during an invited scientific visit. Concomitantly, a style of genomics research arose that emphasizes the accumulation and manipulation of large amounts of data. Will the frontiers of biology soon shift from single investigator academic labs to the top biotech companies, genomics institutes, and large collaborative academic groups?

While some of the problems raised by the success and vast expansion of genetic biology are real, it is clear that these are issues best addressed by a named evolutionary process over the long term. Some steps have already been taken. Several of our facilities were inadequate for important areas we wished to explore. Most glaring was the lack of a modern facility for mouse developmental genetics, a situation that was addressed immediately after I became Director in 1994 by constructing the Keck Laboratory for Vertebrate Genetics on the second floor. Other substantial renovations were accomplished at the same time, including the addition of new laboratories for postdoctoral fellows and for two more staff associates.

We have continuously upgraded our equipment and infrastructure. Since 1994, our computational facilities have expanded enormously as we converted to collecting, analyzing, and presenting all our data in digital form. Now each researcher is provided with a networked computer loaded with a suite of software on their desk, where they seem to carry out an increasing fraction of their research. We have added new equipment, particularly in the area of specialized light microscopes. These instruments, which slice through embryos optically, rather than using physical sections, provide in seconds three-dimensional images that would have taken many days of patient work to re-construct by traditional methods. We have also upgraded the management structure of the Department, which scarcely existed for most of its history. By dividing responsibility for finances, facilities, and information technology among three talented departmental administrators, it has been possible to support the scientific staff better.

During the last few years it became clear that we needed a larger facility to nurture our continued growth and evolution. Many of our scientific resources, particularly our large collections of genetic strains of frogs, zebrafish, *Drosophila*, worms, and mice, as well as modern genomic resources, were filling up what little remained of our unprogrammed space. Having configured every available closet, eventually the halls began to fill with supplies, freezers, and other equipment that would no longer fit anywhere else. We found that it would cost more to renovate our current facility than to build a new one, not to mention the nightmare of moving our complex operation to temporary quarters. Consequently, two years ago with the Institution's support we began to plan for a new departmental facility.

Reaffirming the value of our long-term association with Johns Hopkins University, the Department will relocate to a very attractive site a few hundred yards from our current location. The core of the new facility is a well-lighted central space with an open stairway where people will frequently cross paths on their way to common meeting rooms, shared equipment rooms, and administrative space located nearby. Individual labs lie nearby along three hallways that radiate outward from the center. This design will maintain the small intimate "feel" of the Department, by maximizing natural interactions, while providing space for undisturbed thought and experiment. An enjoyable and critically important aspect of planning the building has been the extensive participation of the entire faculty and their direct interaction with the architects.

Despite a new home and new resources we cannot, of course, escape the long-term forces driving the field of biology toward a "big science" model. Will the field of biology undergo a transition like that in some physical sciences, where collaborative research in large groups becomes the norm? We think this problem has been greatly exaggerated. The field of biology is vast and relatively unexplored compared with the physical sciences. There are tens of thousands of genes in each of tens of millions of species. Although many gene sequences have been determined in a few model organisms we understand only the rudimentary function of just a tiny fraction. We have scarcely begun to understand how chromosome regions are programmed so that an appropriate set of genes become active in each cell. Individual humans contain around 10^{14} cells, comprising thousands of different cell types. Thousands of particular subcomponents reside within each cell type, each a "molecular machine" composed of hundreds of gene products, whose composition, interaction, and functional logic remain mostly unknown. How such a collection of marvelously complicated devices can work harmoniously to form a cell still boggles the imagination. Like trying to comprehend the size of the universe or the depths of time since the big bang, just grasping the full complexity of a cell is extremely difficult and regularly eludes newcomers to biology, who confidently predict that a working mathematical model will soon be forthcoming.

Complex as it is, a cell lies near the bottom on the scale of biological organization. Cells are biology's atoms; their interactions are responsible for most of life's properties and these become exponentially even more complex. Myriad interactions between cells, most of which remain undefined, are critical in multicellular organisms, beginning from the first division of the fertilized zygote. Cell behavior is constantly influenced by the environment, circadian time, nutrition, aging, infections, and even memories. The idea that one simply has to measure which genes are expressed in each cell type to understand an organism is as naive as the claims of the cell modelers. There is no such thing as a fixed gene profile for any cell. At still higher levels, organisms become mere atoms. Organismal interactions give rise to communities, parasitism, disease, ecology, and the evolutionary process that makes it all possible. In the past, a gene-based approach to this vast range of biological phenomena seemed impossible. Now, one could hope that a new way to study many important aspects of biology might be within reach. What genetics has done for developmental biology will be followed by similar revolutions up through the entire scale of the living world.

Thus, the new biological technologies offer some of the greatest opportunities ever available to young biological scientists, rather than a cause for alarm. Despite their rhetoric, the existence of large-scale data gathering operations will simply speed the progress of those individuals clever enough to combine this output with the still essential ingredient of truly novel ideas. Having played a significant role in creating the revolution in biological methods, it has been relatively easy for our faculty to make the mental transitions needed effectively to utilize the rapidly changing methodology.

Where then is the Department's research headed in the twenty-first century? We are proud to say that we do not know. We trust our future will be determined by new, bold, risky ventures on the frontiers of biological science by our most insightful and daring faculty. The opportunities for individual initiative have never been greater. As young staff arrive and as new areas of research are attempted, our path will evolve. One need only throw off the crippling notion that the end of a scientific era is at hand, that we have reached a point in the biological sciences where the basic outlines of living things are known, and that increasingly specialized workers will just dot the i's and cross the t's. The reality is in 2002, as it was in 1902, that nothing could be further from the truth. The Department has succeeded because of its faith in individual creativity, and this will remain our future course.

INDEX

Note: page numbers in *italics* refer to figures and tables. The Carnegie Institution of Washington is abbreviated to CIW and the Station for Experimental Evolution to SEE in subentries.